Advance Praise for *Being the Change*

A plethora of insights about nature and ourselves, revealed by one man's journey as he comes to terms with human exploitation of our planet.

—Dr. James Hansen, climate scientist and former director of
NASA's Goddard Institute for Space Studies

A powerful reminder that it is possible—and joyful—to move away from fossil fuels, even in a society still in the throes of addiction.

— Bill McKibben, author, *Eaarth: Making a Life on a Tough New Planet*

A low-carbon world will not look like *Star Trek*, it will look mostly like it looks today, it's just that we will inhabit it differently. Peter Kalmus's brilliant book is about his deciding to start living that way today. He finds that (a) it's not that hard, and that (b) life improves. He becomes more skilled, connected, fulfilled, nourished. As will we all. Allow him to ease you over the threshold.

— Rob Hopkins, founder of the Transition movement robhopkins.net.

Too many people say that personal action isn't enough to deal with the mess we've made of the global climate, and think that this means personal action isn't necessary. In this timely and provocative book, Peter Kalmus points out that changing the world has to start with changing our own lives. It's a crucial message that needs to be heard.

—John Michael Greer, author of *After Progress* and *The Retro Future*

What does an astrophysicist do if he learns that civilization is on path toward oblivion? If he's Peter Kalmus, he meditates, examines his life, and makes significant changes to reduce his personal carbon output. Then he writes a book. The result is a humane and intelligent exploration of what anyone can do to reduce climate impact—and live a better life in the process.

— Richard Heinberg, Senior Fellow, Post Carbon Institute

A low-emission lifestyle is empowering, happier, and strengthens our connection with community and our environment (plus yes, it saves us thousands of dollars). This is an important and valuable book, and recommended reading for anyone interested in a richer life or a safer climate (doubly so for those interested in both).

—John Cook, research assistant professor at George Mason University
and founder of SkepticalScience.com

Too often, books by scientists err toward the ultra-cerebral. Full of facts, figures and charts—but not enough heart. That's what makes *Being the Change* so refreshing. Kalmus is a respected atmospheric scientist and weighs in with authority when it comes to the topic of climate change. But he speaks to us as a person, sharing his experiences, concerns, and aspirations as a fellow human being combatting the existential threat of human-caused climate change. And he shares with us a vital message about how we can indeed be the change we need to see in the world if we are to avert a climate catastrophe.

— Michael E. Mann, Distinguished Professor of Atmospheric Science, Penn State University, and co-author, *The Madhouse Effect: How Climate Change Denial is Threatening the Planet, Destroying our Politics, and Driving Us Crazy*

When Science and the mind are aligned with the heart, they become True Science and it manifests in books like *Being the Change* — a sort of courageous manifesto for citizens of the World in the 21st century. Peter Kalmus is the kind of dad, husband, friend, serious meditator, scientist, heartivist and brother any of us concerned for future generations and more harmonious communities would like to have in our (Solar)neighborhood.

—Pancho Ramos-Stierle, Satyagrahi and full-time ServiceSpace volunteer

Imagine you had your very own climate scientist living next door. What would he or she tell you to do? Peter is that neighbor. He walks the walk for his kids, for the land, for our future—and he can help you do it too.

— Kelly Coyne and Erik Knutzen, authors, *The Urban Homestead* and *Making It: Radical Home Ec for a Post-Consumer World* rootsimple.com

So often, we feel that nothing we do will make a difference. Peter doesn't just dispel that myth, he buries it: under his feral bee hives, his urban chicken run, and his compost heap (just don't ask what's in it). These gut-wrenchingly honest yet obstinately hopeful reflections provide a roadmap to building our own personal bulwark against the storm we face today.

—Katharine Hayhoe, climate scientist, Professor at Texas Tech University, author, *A Climate for Change: Global Warming Facts for Faith-based Decisions*

This book makes it clear that all of us have a responsibility to cherish the miracles that compose the natural world. We need to think deeply about how we live and then, as Peter Kalmus advises, radically reduce our use of fossil fuels. Everything is sacred. Learning how to get along, to be happy and to live within the limits of the biosphere are sacred tasks. Please read this book. It will be good for your soul.

—James Hoggan, author, *I'm Right and You're an Idiot*

Peter's work makes me smile. The mission of Citizens Climate Lobby is to create the political will for a livable planet by encouraging others to make breakthroughs in their personal and political power. Peter demonstrates practical steps, for individuals and organizations all the way up to the global scale, to advance these goals. His manner of living exemplifies the connection between power, reason, creativity and joy.

—James Waterhouse, co-founder, Citizens Climate Lobby, Pasadena Foothills Chapter; co-founder, SoCal 350 Action Network

We all must take huge risks in order to create a truly just and life-sustaining society. *Being the Change* maps the first important leaps on this journey, describing real-life examples of the good life that awaits us beyond capitalism, species-extinction, economic injustice, and fossil-fuel addiction. It is a roadmap out of our destructive and oppressive culture that touches upon the essential need for wealth redistribution and racial justice in the climate revolution. Please follow Peter's inspiring example: we must act!

—Ethan Hughes, co-founder of the Possibility Alliance

BEING
the
change

LIVE WELL
AND SPARK A
CLIMATE
REVOLUTION

by
PETER KALMUS

new society
PUBLISHERS

Cover design by Diane McIntosh.
Interior illustrations by Sam Bower;
other graphics by Peter Kalmus unless otherwise noted;
p. 1: © oxanaart, p. 3: © Sergey Nivens, p. 125: © hikolaj2 / Adobe Stock.

Printed in Canada. First printing June 2017.

Funded by the Government of Canada	Financé par le gouvernement du Canada	Canada

The ideas and opinions herein are the author's. The author does not speak on behalf
of NASA, the Jet Propulsion Laboratory, or the California Institute of Technology.

The author is donating his profits to groups with potential to connect individual
agency to collective climate action, such as Citizens' Climate Lobby.

Inquiries regarding requests to reprint all or part of *Being the Change*
should be addressed to New Society Publishers at the address below.
To order directly from the publishers, please call toll-free (North America)
1-800-567-6772, or order online at www.newsociety.com

Any other inquiries can be directed by mail to:

New Society Publishers
P.O. Box 189, Gabriola Island, BC V0R 1X0, Canada
(250) 247-9737

LIBRARY AND ARCHIVES CANADA CATALOGUING IN PUBLICATION

Kalmus, Peter, 1974-, author

Being the change : live well and spark a climate revolution / by Peter Kalmus.

Includes bibliographical references and index.
Issued in print and electronic formats.
ISBN 978-0-86571-853-1 (softcover).—ISBN 978-1-55092-648-4 (PDF).—
ISBN 978-1-77142-243-7 (EPUB)

1. Sustainable living.   2. Climatic changes—Prevention—Citizen
participation.   I. Title.

GE196.K35 2017   333.72   C2017-902837-5
                                                        C2017-902838-3

New Society Publishers' mission is to publish books that contribute in fundamental
ways to building an ecologically sustainable and just society, and to do so with the
least possible impact on the environment, in a manner that models this vision.

To Braird and Zane,
and the rest of Earth's children

Contents

Preface

This book explores a lot of territory. It discusses climate science, climate policy, and aquifer depletion—as well as mythology, meditation, and beekeeping. These and other topics herein have been written about in greater detail elsewhere. There are entire volumes devoted to backyard chickens. So why mention them here, in a book with "climate" in its title?

The answer has to do with the nature of our predicament: global warming touches every aspect of our lives. It connects gardening to population growth, bicycling to flying in a plane. Most of all, global warming challenges us to rethink humanity's place in the web of life on this beautiful planet—to reimagine what it means to be human. Global warming is, perhaps first and foremost, a failure of humanity's collective imagination. As such it doesn't fit neatly inside any single box or discipline.

How, then, can we respond as individuals? In searching for answers, I've read books about science, policy, practical action, and spirituality. But none spoke to my being as a whole. And they tended to be too polite, too careful, too narrow; they didn't ask enough of me. Their suggestions were not on a scale commensurate with the scale of the predicament. And far too many were joyless.

As I learned more about climate change, my need to do something intensified. The path was far from clear, but I did my best, gradually and systematically changing my daily life. My response draws on science, practical action, and spiritual examination, and these threads interweave on every level. You hold the unique result in your hands: a book written from the perspective of a meditating climate scientist who has nearly eliminated his own greenhouse gas emissions—and who discovered this to be surprisingly satisfying, empowering, and relevant to collective change. In place of burning fossil fuels, humanity can become smarter, more creative, kinder.

Since beginning down this path, I've covered a lot of ground. I've changed many things about my life and had a lot of fun. At the same time, I've come face-to-face with the seriousness of our climate emergency. To continue business as usual is to tacitly place a blind-faith bet on the emergence of some techno-fix; this amounts to magical thinking. And global warming is happening with a rapidity that leaves me speechless. The longer we take to change direction, the more suffering we'll experience and the longer this suffering will last. And for what? A consumerist lifestyle that doesn't even make us happy. We must do everything we can to change direction. And a big part of this is imagining, living, and telling the stories of *what comes next.*

In addressing something so all-encompassing, you'll ultimately need to forge your own response. My hope is that this book will support and inspire you as you do so.

Writing it has been a long journey, both literally and figuratively, and I'm grateful to the wonderful people who have nourished and sheltered me along the way. These include Audrey, Katie, Christina, my mom and dad, Therese Brummel, Abe de la Houssaye, January Nordman, Lin Griffith, Maya Saran, Baldeep Singh, Paul Livingstone, Mark Rice, John Hopkins, David Sneider, Susan Rudnicki, Paul Taylor, Daniel Suelo, Victoria and Alec Loorz, Russel Greene, Alan Weinstein, Joao Teixeira, Mark Richardson, Ryan Pavlick, Matt Lebsock, Brian Kahn, A.B., Angie Pendergrass, Bryan Allen, Jim Waterhouse, Rob Haw and the rest of the awesome PF-CCL team, Markus Loeffler, Clay Folk, P.J. Parmar, James Bakner, Tera Little, Sarah Baird, Sarah Reber, Ben Denckla, Brent Ranalli, Elizabeth Mathews, Sam Bower, Erik Knutsen, Kelly Coyne and Pancho Ramos-Stierle—many of whom provided detailed comments which greatly improved the manuscript. It was a pleasure working with the artist Sam Bower and the editors Robin Rauzi and Betsy Nuse. I'd like to thank the good folks at New Society Publishers (especially Rob West), and the other good folks at YES! Magazine (especially Tracy Loeffelholz Dunn), for believing in me—and for enabling new voices to take part in this crucial conversation.

Above all I thank my wife, Sharon Kunde, for a lifetime of challenging discussion, insightful comments, unwavering support, and plain old companionship; and for patiently putting up with my many foibles and crazy projects. I couldn't ask for a better Dhamma partner.

— Peter Kalmus
Altadena, California

PART I

PREDICAMENT

Full fathom five thy father lies;
Of his bones are coral made;
Those are pearls that were his eyes:
Nothing of him that doth fade
But doth suffer a sea-change
Into something rich and strange.

—WILLIAM SHAKESPEARE, *The Tempest*

CHAPTER 1

Waking Up

Trees and people used to be the best of friends.
I saw that tree and decided to buy the house.

— HAYAO MIYAZAKI, *My Neighbor Totoro*

I knew that burning fossil fuels was causing irreversible harm to our planet's life-support systems. And yet I continued to burn.

When I first heard of global warming[1] in sixth grade—the only time it was mentioned during my school years—it seemed like science fiction, not something that would ever concern me. I didn't think about it again for nearly two decades.

I began learning the basic science of global warming in 2006 when my first son, Braird, was born. Fatherhood jolted me out of a selfish careerism. Suddenly my life wasn't just about me, and my perspective shifted to a longer time scale. At the time, I was working on my PhD in physics at Columbia University in New York City. As my eyes were opened, I had a strong emotional response: how could we continue burning fossil fuels at an accelerating pace when this severely damages the biosphere for future generations? It seemed insane. At the same time, I was immersed in our industrial civilization, which dictates that burning fossil fuels is the only *sane* thing to do—that someone who *refuses* to burn fossil fuels is ludicrous, a Luddite.

I became obsessed with finding some way to rectify this deep inconsistency. I longed to know how all of the people around me—family members, colleagues, strangers on the street—were dealing with this glaring disconnect without any apparent difficulty. Did they know about global warming? Had they made peace with it somehow, or did they simply not think about it? I felt afraid of the future, lost. I had so much emotional static that I struggled to connect with people.

Like a splinter in my psyche, this disconnect required me to *do something*. But what?

I first tried converting people with facts. The people around me were acting as though there wasn't a problem: perhaps they simply didn't know. If I could only communicate with greater clarity, people would "get it." I felt like I had the truth, that my job was to wake everyone up.

Like most attempts to convert, though, mine were sanctimonious and alienating. It was impossible for anyone to listen to me, or for me to listen to anyone else. (My wife, Sharon, had to put up with a lot; it's not easy being married to someone who wants to convert you.) This led to even more disconnection. Alone with my angst, at a loss for what to do, I was panicking.

I now realize that few people respond to facts. I also realize that I can't respond meaningfully to our predicament with my intellect alone. I also doubt that even our society's collective intellect, our best scientists and brightest policymakers working within their delineated roles, will be enough. While intellect certainly plays a role, it's a rather small one. Our dire ecological crisis calls us to go deeper.

Going deeper

A few years passed before I began to develop a more coherent response. In 2008, our second child, Zane, was born, and we left New York so I could take an astrophysics job at the California Institute of Technology. But before leaving New York, I was offered a job in atmospheric science at NASA's Goddard Institute for Space Studies (GISS), which at the time was led by James Hansen. Had I accepted it, I'd have worked to improve the representation of clouds in the GISS global climate model. But I didn't feel ready for such a big career change, and my ongoing work of searching for gravitational waves—ripples in the fabric of spacetime—was incredibly exciting. So, after much soul searching, I accepted the Caltech job and continued my work of sifting through LIGO data for scientific gold. Sharon and I moved to Altadena, a suburb northeast of Los Angeles in the foothills of the San Gabriel Mountains where parrots roam the skies and orange trees abound. I felt like I'd landed in paradise.

We chose a house because of the magnificent avocado tree in the backyard. I bonded with this tree. I began to think of it as a friend, and I still do. This relationship with a tree began to change me: I began to understand plants as beings.[2]

After a year of renting, we bought the house. For the first time in my life, I owned a tiny patch of land. I decided to cancel the mow-and-blow landscaping service and tend my own yard. The land seemed alien; I didn't know what any of the plants were called or how to take care of them. But I did know that I love to eat tomatoes, so I planted some tomato plants. I enjoyed their company so much—their smell and their just-perceptible daily growth, their being-ness—that I felt called to plant other little beings. I dismantled a small deck by my back fence that we never used, took a sledgehammer to the underlying concrete (quite a joyous task, it turned out), and turned the scrap lumber into six raised beds. I'd caught the gardening bug. Before long I ripped out the grass of my front lawn to make space for other, more interesting and useful plants.

This, then, is how I started to use my hands: the land drew me in. The land was like a painter's canvas, full of possibility and potential. I could plant things on it. Choosing what to grow, and how, required a new kind of wisdom from me, something essentially human. It asked for more than intellect. It asked for connection and for humility, and it offered simple gifts. I fell in love with the land.

I could see a path stretching far into the distance, and I've come to understand that learning how to tend the land takes a lifetime.

Around this time, in 2010, I began to meditate seriously. Sharon and I had started meditating back in New York, but we simply weren't able to maintain our practice while caring for babies. But one morning, after four years of diapers and inadequate sleep, I remembered how important meditation had been. So I went to a ten-day meditation retreat and started practicing again. This is how I started to know myself more deeply. My eyes opened to what was right in front of them. A few months later, Sharon went on her own retreat, and we began sitting together daily.

I began observing my daily life and changing it to be more aligned with what I knew. When faced with some daily task—commuting to work, planning a trip, eating, showering, whatever—I began perceiving how it connects to our industrial system's preferred way of doing things, how it affects other beings and too often harms them. I began searching for alternative ways of doing things. This exploration often blossomed into adventure: unpredictable, fun, and satisfying.

As my scientific interest in global warming increased, it eventually occurred to me that I'd be happier studying it full-time. So I finally left the beautiful, giddy world of astrophysics. This was a sacrifice, and it meant sitting on the sidelines during humanity's first detection of gravitational waves—an endeavor to which I've given nearly a decade of my life. But I simply could no longer concentrate on astrophysics; it felt like fiddling while Rome burned. I'm now an Earth scientist studying the role of clouds in a warming world. I've also reduced my personal CO_2 emissions from about twenty tonnes per year (near the US average) to under two tonnes per year. Overall, this hasn't been a sacrifice. It has made me happier.

Head, hands, and heart

The path I'm on has three parts. One is intellectual understanding: the head. The head allows me to prioritize. It helps me navigate to my goals, although I find it's not always good at *choosing* those goals. One of the lessons I've learned is that I'm limited, in time, energy, and ability; if I'm to make any progress, I need to choose my path wisely. This means asking the right questions, gathering information about reality as it is (which is often different than how it appears to be, or how I want it to be), and drawing conclusions objectively. The head is a scientist.

Another part of my path is practical action: the hands. As we'll see, society's business-as-usual trajectory is carrying us toward disaster. If we wish to avoid disaster, we must take action. Since I can't change the entire global trajectory single-handedly, I perform practical and local actions, changing myself and how I live right here and right now. Direct practical action is empowering; it brings measurable, tangible change. It's fun, and therefore I can sustain it easily. It also provides its own guidance. Time and again I've found that only by taking a step—making some actual change—is the next step revealed. I find that all the planning and intellectualizing in the world can't substitute for just doing something. There's wisdom in doing.

A third part of my path is seeing from the heart. This third part is what connects me to myself, to other people, and to nature. Without it, action can become compulsive, joyless. Connection brings purpose and meaning to thought and action.

I have a specific and concrete practice for this third part: I meditate by observing my body and mind in a particular way. Meditation allows me to be joyful (most of the time) even while studying global warming every day at work. Meditation helps me connect to the sea of everyday miracles around me—the plants growing, the sun shining, my older son lovingly putting his arm around his brother's shoulders. I find great strength in this awareness.

These three parts support and balance one another. In shaping a response to our predicament, each part is important.

Aligning with the biosphere

The changes I've been making to my own life are simple, but they go far beyond recycling or green consumerism. I came to see that the business-as-usual ways of industrial society are bankrupt. So I actively replace those parts of my everyday life that feel unsatisfying with new ways of living that I do find satisfying.

Such changes don't require sacrifice so much as exchange, swapping daily actions that aren't satisfying for ones that are. In this way, my everyday life has gradually come into harmony with my beliefs. My experience has been that congruence between outer and inner life is the key to happiness. I'm no good at fooling myself.

I also came to see how deeply I'd been influenced by the subconscious whisper of culture, how little I questioned my everyday actions, and how completely I accepted the illusion that the way things are is the only way they could be. My old mindset was separation; my emerging mindset is connection. I'm learning that acceptance and detached observation of my own mind is the basis of compassion. I'm learning how to become sustainable, internally.

We could coin a word for this path of inner and outer change: *becycling*, beyond recycling. Becycling entails restoring cyclical natural processes at the local scale. It requires getting busy instead of passively hoping that "they will think of something." It means accepting responsibility for your own everyday actions and changing those that harm other beings in our planet's biosphere. It means actually being the change.

Straightforwardness

My path is straightforward: if fossil fuels cause global warming, and I don't want global warming, then I should reduce my fossil fuel use.

Similarly, if I don't like conflict, killing, and wars, then I should reduce my own addiction to anger and negativity. This seems obvious to me now, but it didn't always. My need to be right used to be blindingly strong, and fear and defensiveness led me to react to anger with more anger, to negativity with more negativity. If we say we want a world without wars, then we shouldn't add hostility to the world ourselves! Yet wherever I go I see people arguing, fighting, and spreading negativity.

In our society, this kind of straightforwardness is often dismissed as idealistic, impractical, and out of reach. But *my own direct experience* says that it is possible to drastically reduce my fossil fuel use, and that it is possible to come out of conflict and negativity. What's more, the personal rewards for doing both are tremendous: a less stressful, more satisfying life.

These two seemingly disparate things—reducing my own fossil fuel use and increasing my ability to love—are actually intimately interconnected. As I learn how to love more, it becomes increasingly clear that I am connected to everything. How, then, can I voluntarily harm the rest of the life on this planet? How can I harm the children who will be born 100 years from now? When someone else suffers, I also suffer. There is no separation between me and the rest of the life on this planet.

To be clear: I'm not saying that selfless love is the near-term answer to global warming. Unfortunately, there are many who, for whatever reason, will never strive to love selflessly; there's no time to wait for them. And even for those who do so strive, it's a long path. This is why we also need sensible policies and technologies that result in cheaper alternatives to fossil fuels.

But for those who are ready to walk on the straightforward path, the path of love, it's certainly worth doing. It may even help to hasten the sensible collective action we desperately need.

Why walk on this path?

I'm aware that the changes I'm making to my daily life will not solve global warming or stave off global economic collapse. How could they? We're rapidly approaching eight billion people on the planet,[3] and I am only one of them.

However, my actions do make me happier, and that's reason enough to do them. I also suspect that, for most of us, individual and local-scale actions are the most skillful means to effect global-scale change. This is a paradox of scale. Our individual actions don't make much of an immediate difference in the global response to our predicament, but they are pieces in a vast puzzle. As more pieces get added, more people will get excited by the emerging picture and begin to add their own pieces.

The prevailing mindset in our industrial society is to search for a silver bullet solution, some brilliant techno-fix that allows us to avoid personal change (which is assumed to be undesirable). After decades of searching by the world's brightest minds, however, it seems likely that there is no such silver bullet. Personal change will therefore likely be necessary. Here are the reasons I'm an early adopter of personal change:

It's enjoyable

In my experience, cutting back on burning fossil fuels became possible—easy, even—when I began to realize that I enjoy my life more when I live mindfully and burn less. I realized that I don't want to burn so much, and I don't need to burn so much. And I genuinely enjoy the changes I've made, such as biking and gardening.

It's empowering

Back when I was concerned about global warming but still burning lots of fossil fuels, I was suffering from cognitive dissonance, living inconsistently. This made me feel depressed and confused. Now I live in a more consistent way, which is empowering. It's the key to connecting with others: my life is my calling card.

I want to help others, not harm them

Burning fossil fuels warms the planet, which harms others. It's that simple. Although the processes involved are distributed globally, accrue over decades, and are statistical in nature—and therefore difficult for our brains to connect directly back to our individual actions—the harm is nonetheless real.

Burning fossil fuels should be unacceptable socially, the way physical assault is unacceptable. The harm it does is less immediate, but just as real.[4] We need to start speaking this truth—burning fossil fuels harms others—so that society can begin realizing it.

It leads to connection and gratitude

Living with less fossil fuels leads to more connection with the land and with my community. It leads to increased awareness that food, water, fuel, and friends are precious. This connection and gratitude makes me happy.

Small actions lead to larger actions

We need to use our unique talents and interests to make a difference, and changing ourselves can reveal how to do this. Small actions gradually led me to two major actions that might have some impact beyond my local community: becoming an Earth scientist and writing this book. These efforts of mine may have larger impact, or they may not. Either way I'll keep making simple changes to my life, while simultaneously looking for opportunities to catalyze collective change.

I've known passionate environmentalists who dreamt of "saving the planet" but who weren't willing to begin changing themselves. But how can we reasonably expect to contribute meaningfully in the larger arena if we can't be bothered to make small changes to our daily lives? If I want to contribute to a change in the narrative, I must begin with myself.

It demonstrates a new story

Few people in the US realize that it's possible to live without fossil fuels. This is a huge failure of imagination. By changing ourselves, we demonstrate what's possible. We explore the new story, and we tell it.

Cynicism and inaction at the national level is nothing more than the collective expression of cynicism and inaction of individuals. When enough of us change ourselves, large-scale change is bound to happen. And when it comes to global warming, our actions speak louder than our words.

It's meaningful

Meaningful work is a great joy. And what could be more meaningful than exploring a new way for humanity to live, in harmony with the biosphere?

As Gandhi wrote: "We but mirror the world. All the tendencies present in the outer world are to be found in the world of our body. If we could change ourselves, the tendencies in the world would also change. As a man changes his own nature, so does the attitude of the world change toward him. This is the divine mystery supreme. A wonderful thing it is and the source of our happiness. *We need not wait to see what others do.*"[5]

Limits, patience, and grief

When I say that I can't save the world, and that I'm aware I have limits, climate activists often misunderstand. They say that I need to stay optimistic, and that I won't inspire anyone by talking about my limits. When they tell me this, I realize that they're operating from one story, and I'm operating from another.

I know that I can change the world; indeed, I am changing the world. What I can't do is save it.

That I have limits is a fact, and I accept it. I don't expect my changes to have a big impact. (I don't expect anything, actually.) If what I do has impact, I know this impact arises only from an existing resonance, a resonance that grows through interacting with many other people in turn. We are like water molecules in a wave: we simultaneously transmit the wave and are moved by it. No one water molecule causes the wave, but together an enormous number of water molecules carry the wave. It's all of us together, carried by a resonance, that will effect great change.

In other words, I operate from the story of the wave, not the story of the hero.

I operate from the story of the wave,
not the story of the hero.

As I ride my bicycle on the overpass over the freeway,
the traffic below looks impermanent.
The way our society lives now feels ephemeral to me.

Sometimes, when I say we need to be patient, activists tell me that the situation is urgent and we have no time for patience. When they say this, I realize they don't know what to do, and that they're panicking. I know because I used to feel that way myself. But in my daily life, patience helps me get tasks done more quickly, not less quickly. Why would responding to global warming be any different? Patience is usually the fastest way to get somewhere worth going to.

I'm aware of how serious our predicament is. I've gone through a process of grief. My grief was deep and intense. It felt like I was part of the ocean, like I was connected to everything. Every now and then this grief comes back to remind me why I do what I do. It purifies and clarifies. I doubt that anyone who understands the seriousness of global warming can avoid this grief.

However, this grief is very far from despair. Grief comes from love, while despair comes from fear. I don't despair; instead I feel joy. It's true that we've lost a lot—a lot of wondrous species, a lot of beautiful places, a lot of opportunities—and that we'll lose even more. But even through this loss, we can experience how much there is to love, how much there is left to save. Our grief and love can lead us to move forward with more creativity and more joy than we ever thought possible.

I have no blind hope that "they will think of something," and yet I still feel optimistic in my own way. My particular optimism comes from the direct experience of connection.

Lifting the illusion

It used to be difficult for me to imagine living in another way, to imagine this land around me in Southern California without freeways, parking lots, or gas stations; to imagine the world without the constant noise of cars, helicopters, planes, and leaf blowers. These seemed like permanent fixtures. I took the conveniences of modern industrial life for granted—the frozen foods aisle, the cheap airplane flights, the internet, the constant distractions. I was attached to them; I wanted more of them. I kept hoping that more would make me happy. More stuff, more money, more clickbait, more convenience. After all, that's what our culture of industrial civilization—*petroculture*—constantly whispers to our subconscious: more of this and you will finally be happy.

Now, as I ride my bicycle on the overpass over the freeway, the traffic below looks impermanent. The way our society lives now feels ephemeral to me.

While I used to see the future as more, I now see it as less. Far from feeling scary to me, less feels right. I've learned that wanting more actually gets in the way of happiness. The feeling of "more and then I'll finally be happy" is an illusion.

I now see the imminent transformation of all that's around me not as an end but as a beginning. This shift in my way of thinking has grown over time out of many moments of simple connection to nature and to other people. Even on a warmer planet, even after today's global industrial civilization is no more than legend, there will still be mountains and sunsets, forests to walk in and oceans to sail, and good people to enjoy it all with.

But there's lots of work to do to prepare for the coming storms. Happily, the work is fun.

CHAPTER 2

Beyond *Green*

If we feel helpless or overwhelmed,
if we have anger, fear, or despair, then no matter
what we do to heal ourselves or the planet,
it will not succeed.

— THICH NHAT HANH, *Love Letter to the Earth*

Language both reflects and shapes how we see the world. The words we use to talk about our predicament reveal fundamental assumptions in how we perceive and relate to nature. Taking words for granted leads to confusion, and worse.

In this chapter, I discuss a few words and concepts we may be better off abandoning, and suggest some alternatives. In doing so, I hope to provide insights into some limitations of current environmental thinking, and to develop a new mindset that will better serve us as we revise humanity's relationship with the biosphere.

Nature and environmentalism

The word *environment* (as typically used by environmentalists) implies a dualism, a competition between the needs of humans and the needs of a nonhuman environment. It has become interchangeable with the word *nature*, which no longer signifies the totality of the physical universe, but instead signifies the domain of nonhumans. This dualism contributes to human exceptionalism, the idea that humans are outside of nature, unbound by natural laws, special among all species.

The reality, though, is that we're one among millions of species supporting each other (while simultaneously competing with each other) in the diverse web of relationships that is the biosphere. The human species depends on this biosphere just like every other species on Earth. The biosphere gives us food, water, oxygen, and a climate in which we can survive. At this level of understanding, there is no dualism. We are nature, and nature is us.

The dualism in the word *environment* manifests on the left as the idea that the environment needs to be saved, and on the right as the idea that the environment is humanity's to extract and exploit. These worldviews are actually two sides of the same coin, stemming as they do from a false sense of separation and human exceptionalism.

Biospherism

When we talk about the environment, we're usually talking about the biosphere or some part of the biosphere. Why not just say "biosphere"?

Whereas environmentalism seeks to protect the environment from humans, *biospherism* seeks to transition to a way of life that respects the limits of the biosphere and all life.

> What's at issue in fact is not an environment; it's a living world.
>
> — DAVID QUAMMEN

Whereas environmentalism implies duality, biospherism implies unity. Whereas environmentalism is reactive, chasing after the latest disaster, biospherism is proactive, seeking to transform the way we think and live. Whereas environmentalism treats the symptoms, biospherism treats the underlying cause.

Humans will always have an impact on the biosphere, and biospherism doesn't seek to eliminate our impact. Biospherism accepts that the biosphere just is the sum of the impacts of individuals (human and nonhuman, from any of the kingdoms of life) comprising it. It seeks to reduce human impact to sustainable levels by changing our priorities.

Biospherism seeks balance. It's the word I'll use in place of *environmentalism*. Someday I hope we can drop such terms altogether and simply say that we're human, and it will mean we live aligned with the biosphere, with each other, and with ourselves.

Beyond fear-and-guilt environmentalism

Environmentalism has had a strong tendency to use shame, guilt, and fear in an attempt to motivate action. But guilt and fear don't motivate me—they discourage me.

It's common for mainstream environmental speakers and writers to put a long and fearsome litany of climate change consequences front and center.[1] These presenters assume their audiences aren't aware of how scary global warming is (because if they were, the assumption goes, they'd certainly act). They therefore communicate fear with visions of hellfire and brimstone. At the end they tack on a few superficial suggestions, "ten things you can do" such as changing light bulbs or shopping at farmers' markets. Finally, they add a thin veneer of hope: "there's still time, but we must act now."

Hellfire and brimstone don't inspire us to change; they lead to guilt. Guilt is a coping mechanism that allows us to merely limp along with our anxiety. It's what we feel when we engage in some action that goes against our deeper principles, but that we don't actually intend to change. Guilt is an insincere self-apology for a painful internal fracture. It leads us to symbolic actions that allow us to function with this fracture. Why not just heal the fracture?

Interestingly, some of the most prominent leaders in the environmental movement reveal this inconsistency between their actions and their edicts. They tell us to stop burning fossil fuels, and yet they themselves have outsized carbon footprints. This hypocrisy might help to explain why the movement itself swirls with guilt. It may also help to explain why it has been ineffective. I suspect that most people notice hypocrisy at some level, and that it has a paralyzing effect. People think, "If even prominent environmental leaders can't reduce their carbon footprints, then it must be impossible."

What an incredible thing, a miraculous thing—
and now we know, a harmful thing—
to fly in an airplane.

While we do need to change ourselves, we also need to forgive ourselves. Those of us who were born into industrial society entered a powerful system that determines our beliefs and daily actions. Socialization colors how we see the world and makes it difficult, maybe even impossible, to see objectively. For example, until recently I drove cars and flew in airplanes without realizing their harmful consequences. Isn't it remarkable that as a society we take flying in airplanes for granted? What an incredible thing, a miraculous thing—and now we know, a harmful thing—to fly in an airplane.

It's time to move on to a more mature advocacy focused on developing a vastly deeper response to the predicament we face, beyond recycling and shopping for "green" cars and carbon offsets. Let's instead learn how to live in alignment with the biosphere, both as individuals and as a collective. This practice demands that we change our everyday lives, how we think about ourselves and our place on this planet.

Earth is a wondrously beautiful place, and will remain so even as we pass through this ecological crisis and ultimately come out the other side. Let's not fear our mother when she's sick. Instead, let's learn to feel compassion for her, and remind ourselves how precious her gifts are. Let's cultivate fierce and fearless love. And for goodness sake, let's stop performing the daily actions that are sickening her! We can stop burning fossil fuels out of a sense of compassionate love. *This* is the action we must perform. The second part of this book is about *how* to do this.

Let's not go green

The word *green* has been thoroughly co-opted by corporate marketing. Maybe it was useful once, maybe not, but now it zombie-walks through environmental discourse.

The word has no precise meaning in an environmental context, yet it strongly signifies vague environmental virtue. This makes it the perfect word for corporations seeking to profit from environmental guilt: "Go green! Buy our product (and feel better about yourself)." The corporations even get to decide what counts as green; in the US there's no regulation of green advertising. Corporations that do great damage to the biosphere regularly brand themselves as green, including car makers, airlines, and fossil fuel producers. It sometimes seems as if the more damaging a corporation is, the greener it claims to be.

Buying green stuff promotes the *status quo* consumer mindset. Green allows us to feel like we're responding to our predicament without needing to change. Green precludes meaningful action, and in this way does more harm than good. Our predicament is deep, and it demands a deeper response from us than shopping.

Low-energy

I propose *low-energy* as a replacement for *green*.

Using less energy at the global scale would reduce greenhouse gas emissions and serve as a bridge to a future without fossil fuels. Using less energy in our individual lives would equip us with the mindset, the skills, and the systems we'll each need in this post-fossil-fuel world.

If the adjective *low-energy* replaced *green*, its specificity would encourage meaningful collective action, such as using less energy. Furthermore, it could not be co-opted. Low-energy could not be used to sell airplane flights, air conditioners, or other fixtures of a high-energy lifestyle.

Many of the changes I've made to my daily life originated from realizing how precious energy really is. I think most people are afraid of a low-energy lifestyle because we equate quality of life with quantity of energy use. My experience has been the opposite: low-energy living is more fun and satisfying.

Sustainable and regenerative

The word *sustainable* is everywhere, but what does it actually mean? The literal meaning is "able to endure." Sustainability therefore involves both a time scale and an object: something is sustained, for some length of time. Thinking about sustainability, then, means thinking about change. This makes it clear that nothing sustains forever.

When we talk about sustainability, we're usually talking about a way of living, a relationship between humans and the biosphere. What time scale should we choose? We need a time scale that reflects the changing biosphere. One hundred years is too short, only a couple of human generations. Fifty thousand years is too long: there's already evolutionary change on this time scale. Indeed, we evolved to become cognitively human only 50,000 years ago. I suggest we aim for a way of living that we can sustain for 1,000–10,000 years. We can use this working definition to evaluate specific human behaviors.

Exponential population growth at a rate of 1.7% (the long-term historical rate; see Chapter 4) is no longer sustainable. After 1,000 years, at this rate we'd have 176 million billion people—which works out to 1,200 people on every square meter. We humans wouldn't even fit on the planet. So our growth will necessarily change—and, in fact, it is changing. Roughly speaking, having more than two children is not currently sustainable for our planet.

Our path to long-term sustainability is to stop growing and to find balance: to pull back to a global consumption and population that the biosphere can sustain. This will require a deep cultural shift, especially within affluent societies and minds. And if we don't make this change, the biosphere will do it for us, for example through global warming-induced disease or famine.

We can go a step further and think in terms of *regeneration* rather than sustainability. Doing so neatly sidesteps the need for a time scale, and it embraces the concept of change. Regeneration means bringing some part of the Earth, or some part of the human way of life, back into alignment with the biosphere. Regeneration calls us to do more than merely sustain: it calls us to heal, and to make our lives expressions of love for all beings.

What would a regenerative society look like in practice? For starters, it would respect the regeneration rate of every resource. Its food system would not depend on fossil fuels, and regions using groundwater would do so at a rate less than the aquifer's regeneration rate. Energy use would be essentially limited to what we could glean from the sun and wind. Metals would be entirely recycled. The population size would remain steady at a biospherically appropriate level, and economies wouldn't depend on growth. Huge swaths of land and ocean would be allowed to rewild. Science and technology would continue to thrive, but their focus would shift: science might be more interested in understanding the relationship between fungi and plants and might no longer concentrate capital for ever-larger atom smashers; technology might focus on doing more with less. A regenerative society would necessarily be more just and equitable. Accumulating wealth would no longer be the main goal of life.

Whether humans are capable of this transformation or not remains an open question. But changing yourself is one way to vote for it.

Recycling

Somewhere within our industrial mindset, there's a place called Away. When something breaks, or bores us, we throw it in the garbage and trash collectors take it Away. We flush a toilet and invisible pipes take it all Away. However, we are slowly learning that Away was always really just Somewhere Else, because everything is connected. But despite our increasing awareness, most of us still haul our bins to the curb and flush our toilets. It feels like we have no other choice. (There are other choices. See Chapters 12 and 13.)

This explains why industrial society fetishizes recycling. Recycling seems like a good thing on the surface, but it contributes to the broken *status quo*. Doesn't recycling help to keep the concept of Away alive in some sense? I know it does for me. I throw a plastic bottle into the recycling and I like knowing it goes Away—but to some better Away. Recycling helps me feel good about Away and allows me to go on consuming as before.

I'm not saying we shouldn't recycle. I'm saying that we shouldn't let recycling stunt our awareness of the impacts of our consumption. Recycling is Garbage 2.0. Let's reduce what goes in the recycling bin, as well as what goes in the garbage bin.

Independence, self-reliance, community-reliance

Independence is an illusion. If you truly depended on nothing, it would mean that you could float out in deep space by yourself, alive and happy. We certainly depend on our biosphere. We also depend on each other. If you depend on some tool for survival, a parka or a knife perhaps, doesn't this mean you depend on the people who made that tool, and the people who made it possible for those people to make the tool? And would life be meaningful if it were lived in isolation, apart from any other person?

Self-reliance differs from independence. I'm self-reliant when I rely on myself first. Ironically perhaps, self-reliance can make an individual a more valuable member of the community. A self-reliant person can solve problems and find new ways of doing things; has a wide array of skills; is confident and optimistic; is strong and able to help others.

In my experience, *community-reliance* grows out of self-reliance. Community-reliance means contributing to community, so that the community is strong and there for you when you need it.

I reject selfish survivalism, heading to the hills with guns and a supply of food. While I do think we need to first look to ourselves for our security (self-reliance), we need to do this within the context of community. Selfish survivalism is ultimately a losing strategy.[2]

Problem, predicament, challenge

I used to think that climate change, overpopulation, and biospheric degradation were *problems*. In identifying them as problems, I assumed there were solutions, which kept me from seeing that my way of life had to change. I really believed that the future

> A civilization that tried to turn all its predicaments into problems has been confronted with problems that, ignored too long, have turned into predicaments.
>
> — JOHN MICHAEL GREER

would look like *Star Trek*, a comforting belief. Perhaps there were solutions a few decades ago. For example, we could have avoided climate change if we'd started seriously addressing it in 1986, the year Ronald Reagan ordered the solar panels on the White House roof be taken down.[3]

At this point, though, we can't avoid climate change for the simple reason that it's already here. Global surface temperatures have already increased by more than one degree Celsius, and additional warming is guaranteed no matter how quickly we reduce our fossil fuel use. What was a problem with a solution in 1986 has become a *predicament*. We probably can't solve it, but we can choose how we respond to it and how bad we let it get.

A predicament is an existential challenge. We cannot make it go away. Death, the archetypal predicament, challenges us to respond by finding meaning in our brief lives. Likewise, I think our collective socio-ecologic predicament challenges us to find out who we really are and what it means to be children of this Earth, in harmony with ourselves, each other, and the rest of the biosphere.

Re-minding

I doubt we'll come through our predicament without a deep change of mindset, a kind of rebirth of our shared existential worldview. Maybe this change will originate within us, or maybe the change will originate externally, catalyzed by the disasters we are bound to experience as our predicament deepens. Either way, we will be re-minded of what is important.

The energy that changes mindset from within is *mindfulness*. Mindfulness means every moment awareness—being aware of reality as it is, as manifest in the mind and the body, from moment to moment. When mindful, I'm present for the reality of this moment, not rolling in thoughts of the past or the future, or wishing for something other than what is. I'm aware of the action I'm engaged in and its consequences, not acting on autopilot; and this awareness of the present moment and its consequences is what drives self-change.

However, in my experience it's not possible to simply decide to "be mindful." Developing mindfulness takes dedicated practice, as I'll discuss in Chapter 11.

Happiness

When I experience some success, my mind is excited and full of a pleasant sensation. I feel larger, like there's more of "me." I've learned that this feeling isn't happiness. Rather, it's the ego being inflated.

I think that we often mistake this sort of ego-excitation for happiness. This is a mistake: it causes us to chase after things that ultimately increase our suffering. Real happiness doesn't depend on external situations. Instead, it's a sense of peace and wellness, of satisfaction and wholeness, a sense that it's wonderful to be alive, a joy in the happiness of others. Real happiness has no anxiety or craving. It vibrates with gratitude, and translates into an eagerness to help others, to spread happiness. Unlike ego-excitation, which is directed toward the self ("I win!"), real happiness is directed toward others, and all of life ("It's a miracle to walk on this Earth!").

As I become happier, the roller-coaster ride of my ego becomes less wild. The lows become less severe: when I fail I find myself smiling with kind laughter, as with a child who is learning to walk. The highs become opportunities to serve. I ask myself "how can this success help others?"

Saving the planet, saving the world

"Saving the planet" is a fantasy for society's collective ego. It allows us to continue in our false belief that we're separate from the biosphere, that what's happening to "the planet," while sad for polar bears, somehow won't affect us.

If you feel discouraged, maybe you're trying to save the world. It's discouraging to have an impossible goal. I think there are a lot of people who subconsciously want to save the world. But saving or not saving is a false binary, and arises from the same instant gratification mindset that got us into this predicament in the first place. Saving the world is a fantasy for our egos.

The opposite of wanting to save the world is having sincere patience. With patience comes humility, openness, and a more skillful capacity for positive change.

> How dare you talk of helping the world? God alone can do that. First you must be made free from all sense of self; then the Divine Mother will give you a task to do.
>
> — RAMAKRISHNA

We each have the power to make the world better, or worse. Each of us can choose to push the world toward a warmer temperature, or pull back. I used to want to save the world. I've finally accepted that I can't, and this has brought me peace. Instead, I try to live a good life so that I can *change* the world.

CHAPTER 3

Global Warming: The Science

*These are the kinds of disciplines in the field of science
that you have to learn—to know when you know
and when you don't know, and what it is you know
and what it is you don't know. You've got to be
very careful not to confuse yourself.*

— RICHARD FEYNMAN

Human influence on the climate is now crystal clear. But the general public has so far failed to understand how rapidly humans are warming the planet and how irreversible the changes will be.

The rapid progression of global warming continues to amaze me. Our CO_2 emissions, which drive modern-day global warming, are following an exponential trajectory. Many changes in the Earth system are accelerating. It's fair to say that scientists, as a group, are surprised by the rapidity.

In this chapter and the next, I hope to clarify two basic time scales of our predicament: speed of onset and duration. I also hope that this brief tour of Earth science will enrich you with a deeper understanding of your relationship with this beautiful planet.

I haven't attempted to write a mini-textbook or to be complete. (An attempt at completeness, the Intergovernmental Panel on Climate Change's Fifth Assessment Report[1] weighed in at 4,852 pages.) Neither have I attempted to describe past climate changes. I've attempted to give the context necessary for a basic understanding of the global warming that's happening at this moment, but not to overwhelm you with more than that.[2]

At the topmost level, climate science has one thing to teach in regards to the well-being of our species and the rest of the biosphere: *to curtail global warming, stop burning fossil fuels*. Of course, there's no guarantee that humans will stop burning fossil fuels in time to avoid truly catastrophic warming. Indeed, part of the scientific work is in understanding how the Earth is likely to change at different emission levels. This is an important discussion for us to have, calmly and with vigilant attention to the evidence, as we boldly continue up the exponential curve into this unprecedented age of planetary change.[3]

The weight of this knowledge

Before diving in, it's worth acknowledging that learning about global warming can be stressful. I once had a friend tell me that she didn't want to know about global warming because she was afraid of becoming too anxious or depressed to function effectively. This is a precarious and short-term stance to take, however; the evidence of global warming will continue to mount in our daily experience, and therefore the psychological stress of holding its reality at bay will also mount. I personally find it less stressful to face the reality of global warming and to begin responding appropriately.

As I described in Chapter 1, though, I had to go through a grieving process to get to this point. And nine out of ten of my colleagues in Earth science have also grieved to some extent—though you'd never know this unless you performed an anonymous survey (as I did).[4] I think the scientific community's response to global warming has been objective to a fault. We're scientists, yes, but we're also humans. If we'd let our humanity shine out more, perhaps it would help our message get through.

The year of climate departure

The first thing to know about global warming is simply that it's already here. The global mean surface temperature rose by 1.0°C between 1880 and 2012,[5] and many impacts of this warming are already clear. The second thing to know, perhaps, is how fast it's progressing.

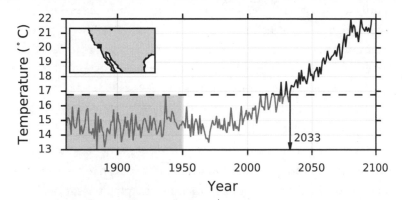

FIGURE 3.1. The predicted climate departure of the grid point containing Altadena, California (the small square on the map) under the business-as-usual scenario (RCP 8.5), for a single climate model (called HadGEM2-ES). The historical baseline variability is shown in grey shading. The arrow indicates the year of climate departure. Different models give different estimates of departure.

We can ask a simple question: at a given location, when will the annual mean temperature exceed the hottest year from a historical baseline, never to return? After this year of no return, the climate at that location will be in a new regime; the climate will have "departed." Figure 3.1 illustrates this.

Camilo Mora and his colleagues at the University of Hawaii explored this question,[6] using estimates of surface temperature from 17 separate global climate models simulating the planet from 1860 until 2100, for over five million 100 km by 100 km grid cells on the planet.[7] They analyzed model runs simulating two global emission scenarios. In one scenario, humanity makes only a modest effort to reduce emissions, which continue to grow until 2100 and beyond (business-as-usual, i.e., what we're doing now). In the other scenario, humanity makes a stronger mitigation effort such that emissions peak near the year 2040 and then decline.[8]

Each region on the planet has its own predicted year of annual climate departure for a given scenario. After departure, a region will still occasionally experience cool days or even cool months, relative to its historic climate; but there will be no cool years there until after the age of global warming. Under the business-as-usual scenario, Mora et al. estimate that global average climate departure (the average over regional departures) will occur in 2036, less than 20 years from now. Under the mitigation scenario, global departure is delayed by an estimated 15 years, to 2051.[9]

Departure will occur first in the tropics, since there is less year-to-year variation there. This is unfortunate both for people in developing nations, who have contributed the least to global warming, and to species in biodiversity hot spots like the Amazon rain forest, where plants and animals are adapted for survival in a narrow range of temperatures.

Global climate departure is no longer avoidable, and it's coming very soon.[10] It will likely be here while my children are still in their twenties. Whether this rapidity of the onset of global warming is bad or not is a separate question, and depends on one's values.

Generally speaking, a given plant, animal, or human civilization is evolutionarily adapted to a specific range of temperature, precipitation, and other climate conditions. If the climate moves out of this range, the plant, animal, or civilization must move, adapt, or die. These migrations, adaptations, and deaths are already causing disruption for both humans and nonhumans. My opinion is that this disruption clearly outweighs any benefits from warming. We'll take a brief look at impacts in Chapter 4.

Peak temperature: Why mitigation is crucial

The proximity of climate departure is disturbing, but in my opinion it's worth working toward mitigation no matter how late it gets or how warm it becomes.

This is because peak temperature is still up to us. No matter the date of climate departure, global warming likely will trace a trajectory in which the climate warms, reaches a peak temperature, and then gradually cools down over many millennia. But *how much* it warms will make a difference, largely determining the long-term (e.g., after 2100) depth of impacts such as crop loss, sea level rise, precipitation changes, ice loss, heat waves, and loss of species.[11]

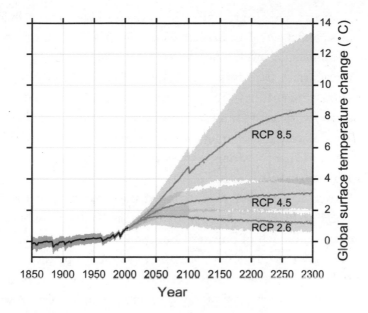

FIGURE 3.2. Temperature trajectories relative to the 1850 to 1900 mean, for three CMIP5 RCP experiments. Solid lines give the multi-model mean of the global mean surface air temperature; shading gives the 5 to 95% uncertainty range across the model ensemble (that is, 90% of model predictions lie in the shading). The black line gives the multi-model mean for the historical experiment (with results from 1850 to 2005). The discontinuity at 2100 has no physical meaning; it's due to the fact that not all models ran to 2300.

Temperature trajectories (with large uncertainties) for various *representative concentration pathways* (RCPs) from now to 2300 are shown in Figure 3.2.[12] RCPs define future greenhouse gas concentrations under hypothetical emission scenarios. Scientists can then run global climate models with these predefined concentrations, facilitating comparison and collaboration. RCP 8.5 is the business-as-usual scenario, while the other two RCPs in Figure 3.2 represent varying levels of mitigation. The lower the RCP number, the more mitigation it presumes.[13]

In RCP 2.6, greenhouse gas emissions peak before 2020 and then decline rapidly, and warming stays below 2°C. However, this pathway is no longer achievable due to our collective procrastination.[14]

RCP 4.5 is a less aggressive pathway that's still open to us. In it, models predict global mean surface warming of 2.4±0.5°C by 2100 over preindustrial levels (and 3.1±0.6°C by 2300).[15]

Our current trajectory is best approximated by RCP 8.5. If we choose this pathway, models predict mean surface warming of 4.3±0.7°C by 2100 (and 8.4±2.9°C by 2300). Note also that in RCP 8.5, warming accelerates rapidly in the 21st century: global temperature rises more during the second half of the century than during the first half.

To put these numbers in one context, the last glacial maximum, in which ice sheets covered not only Greenland but most of North America and much of northern Europe and Asia about 20,000 years ago, was 4.0±0.8°C cooler than in modern preindustrial times.[16] To put them in another context, the last time the planet was about 2°C warmer was 125,000 years ago (the Eemian Maximum, long before humans evolved),[17] and it has been tens of millions of years since the planet was 4°C to 8°C warmer.[18]

The distinction between social and scientific uncertainty is well-illustrated by Figure 3.2. The large spread covered by all the different RCPs indicates social uncertainty; physical science cannot predict what humans will do. On the other hand, the width of the shaded swath for each RCP gives a particular estimate of the scientific uncertainty: the spread between different models.[19] This estimate, however, doesn't capture uncertainty due to physical processes that aren't modeled in the first place—the "unknown unknowns." These could include various carbon cycle feedbacks, discussed below, which might augment warming.

Under all of the RCP scenarios, significant warming persists for many centuries, far beyond 2300, but the peak temperatures are very different. It's unclear exactly what those peak temperatures will be or when they'll occur because far-future model predictions become increasingly uncertain. But I think for our purposes here, and for the purposes of policymakers, Figure 3.2 provides a clear enough picture: to prevent warming far beyond

levels humanity has ever experienced, we need to mitigate immediately and rapidly.

The physical basis for warming

Let's now examine the causes of warming and how warming interacts with the Earth system. Along the way we'll continue developing our understanding of what we know and what we don't know.

The Earth system

The Earth system is tremendously complex. Its main parts, viewed
in the big picture and from the perspective of climate, are the atmosphere, the ocean, the land, the ice, and the biosphere. They
interact with one another via physical, chemical, and biological
processes, over space scales ranging from microscopic to planetary, and time scales ranging from nearly instantaneous to billions
of years.

Before I switched into atmospheric science, I studied neutron
stars and black holes. Everything knowable about a stable isolated
black hole is (we think) encoded in just three numbers: mass, electric charge, and spin. But the state of the Earth depends on every
cloud, tree, drop of moisture, mountain peak, ocean eddy, patch
of snow, bacterium, and internal combustion engine. Instead of
just three numbers, the state of the Earth system is described by
an essentially infinite number of numbers. And to understand the
system, we need to understand how they interact.

Despite this internal complexity, the Earth's climate system
interacts energetically with the universe in only three significant
ways: by absorbing sunlight, by reflecting sunlight, and by emitting
infrared light. Infrared light is invisible to humans, but we feel it
as radiant heat when we sit near a fire. The key fact about infrared
emission is that *hotter objects emit more infrared energy than cooler
objects*. This fact allows the Earth system to balance the sunlight
energy coming in with infrared energy going out to cold space.
For example, if the sun became dimmer, the Earth would cool.

A cooler Earth would emit less infrared light, eventually arriving at a new balance at a cooler temperature.

The greenhouse effect

Greenhouse gases like CO_2 act like a blanket warming the planet. We need this blanket. Without it the Earth's average surface temperature would be $-18°C$ ($0°F$) and there could be no life as we know it.[20] So the greenhouse effect per se is not a bad thing. The problem is that by burning fossil fuels into the atmosphere we're causing the blanket to become warmer. By 2014 we'd already increased the atmospheric CO_2 fraction by 43% over preindustrial levels, and this increase is accelerating exponentially.[21]

Have you ever thought about how a blanket works? Imagine being naked without a blanket on a cool, windless night. Like any warm object, your body emits infrared radiation which carries away energy, cooling you off. Now imagine that you have a blanket. The underside of the blanket absorbs your emitted infrared energy, heats up, and then re-emits infrared radiation back to you. Some heat, however, is conducted through the blanket, causing the relatively cool top side to emit more infrared energy out into the air. *But the top side of the blanket is cooler than your body*, so it radiates less energy. This system reaches equilibrium when the top of the blanket loses infrared energy at the same rate that heat is conducted from its underside. The thicker the blanket, the less heat it conducts, and the hotter it gets underneath before reaching equilibrium.[22]

In place of body heat, the Earth's main source of energy is incoming sunlight. About 70% is absorbed, and the rest is reflected back into space by clouds, ice, snow, and other bright surfaces. Like your body on a cold night, the warm Earth loses heat by emitting infrared radiation into space. Greenhouse gases in the atmosphere act like the blanket, with a warm underside facing Earth (the lower atmosphere) and a colder top side facing space (the upper atmosphere). Greenhouse gases in the atmosphere absorb some of the infrared energy emitted by the Earth's surface.[23] The warmer lower atmosphere then radiates some of this infrared energy back down to the Earth's surface. It radiates in the upward direction, as well; this radiation is trapped by the higher atmospheric layers, in turn (think of the atmosphere as many thin layers). The upward-directed infrared radiation from the cold upper layer streams into space, but since that highest layer is colder, it emits less energy than the planet's surface.

What if we suddenly increase infrared-absorbing greenhouse gases? This makes the atmosphere act like a better blanket, and a smaller fraction of the upwelling surface infrared escapes into space. Because the absorbed solar energy hasn't decreased,[24] there is now an energy imbalance, and Earth warms. Warmer objects emit more infrared radiation, and eventually the escaping infrared energy will once again balance the incoming solar energy, despite the warmer blanket. Eventually the Earth will regain energy balance *but at a warmer temperature.*

Greenhouse gases

The main two human-emitted greenhouse gases are carbon dioxide (CO_2) and methane (CH_4). Human emissions of halocarbons and nitrous oxide (N_2O) also contribute, but to a lesser degree. Each gas is made from atoms electromagnetically connected in a particular geometric configuration. These geometries have specific resonant frequencies that determine how the gas interacts with infrared radiation.

The Earth system exchanges energy primarily by absorbing shortwave solar radiation and emitting longwave infrared radiation. We call factors that change one or the other of these two quantities *radiative forcings*. The Earth is in energy balance when the net radiative forcing is zero. Increasing the atmospheric concentration of a greenhouse gas decreases the outgoing longwave radiation, and this change (in units of power per area, W/m^2) is an example of a radiative forcing.

Water vapor (H_2O) is the largest contributor to the greenhouse effect, but we humans have no direct control over it. It remains in dynamic equilibrium, evaporating into the atmosphere and condensing out as rain. A hotter atmosphere, though, holds more water than a cooler atmosphere. As we warm the atmosphere with the other greenhouse gases, water vapor acts as an amplifier. Ozone (O_3) is another greenhouse gas which humans influence indirectly (via atmospheric chemistry).

Today human emissions do directly influence the atmospheric amounts of the other greenhouse gases. The global warming impact from emitting a tonne of some greenhouse gas depends on how efficiently that gas absorbs infrared light, as well as how long it stays in the atmosphere—its *residence time*.[25]

To allow for apples-to-apples comparisons of different greenhouse gases, we can integrate atmospheric absorption over the residence time of a gas species to calculate its *global warming potential* (GWP). GWPs are estimated relative to CO_2, and given in units of "CO_2-equivalents" (CO_2e). For example, after 100 years, a tonne of methane causes a total of about 34 times more warming than a tonne of CO_2; we say it has a GWP of 34 on a 100-year horizon, or GWP_{100} of 34. However, methane is reactive and has a residence time of only about 12 years, so on a 20-year horizon, its total warming potential relative to CO_2 is even higher: the GWP_{20} of methane is about 105.[26]

The choice of time horizon is subjective, but important—especially for methane. Some methane inevitably escapes during natural gas extraction, processing, and distribution. Most analysts choose GWP_{100} over GWP_{20}, downplaying the contribution of this leakage to global warming and making natural gas appear more attractive as a "bridge fuel."

Nitrous oxide, N_2O

Anthropogenic nitrous oxide in the atmosphere is produced mainly by the agricultural use of nitrogen fertilizers. It's also produced by internal combustion engines and the breakdown of livestock manure and urine. It resides in the atmosphere for 120 years, and has GWP_{20} of 260 (with a GWP_{100} that's essentially the same due to a residence time of over 100 years).[27] Human emissions of nitrous oxide accounted for about 5% of the current greenhouse radiative forcing (measured in 2011; see Figure 3.7, page 50, which we will discuss below).[28]

Halocarbons

Halocarbons are chemicals containing at least one carbon atom and halogen atom (usually chlorine or fluorine), useful as refrigerants, solvents, pesticides, and electrical insulators. They were regulated in the 1990s because they deplete Earth's protective stratospheric ozone layer (and also create dramatic "ozone holes" over the poles). Out of this family of compounds, the CFC-12 (CCl_3F, brand name Freon-12, formerly common in refrigerators, Silly String, air horns, gas dusters, and other applications requiring an easily compressible gas) still has the most impact on the climate, with a residence time of about 100 years and a GWP_{20} of about 10,800.[29] While emissions of CFC-12 have stopped, its global warming impact will continue for many decades. Meanwhile, emissions of other halocarbon compounds are increasing. Human emissions of halocarbons account for another 5% of the greenhouse radiative forcing (see Figure 3.7, page 50).[30]

Methane, CH_4

Methane is a powerful greenhouse gas (GWP_{20} of about 105,[31] with an uncertainty of about 30%) with a short residence time of about 12 years. These two facts mean that mitigating methane emissions would have an immediate and significant impact on our warming trajectory.

Human emissions of methane account for about 30% of the current (instantaneous) greenhouse radiative forcing (see Figure 3.7, page 50).[32] In terms of GWP (integrated over time), in 2010 it accounted for 16%[33] or 37%[34] of anthropogenic greenhouse gas emissions on a GWP_{100} basis or a GWP_{20} basis, respectively.

Over the last 200 years, atmospheric methane concentration has almost tripled, from 650 ppb (parts per billion) to 1,800 ppb.[35]

TABLE 3.1. Sources of anthropogenic global methane emissions, in megatonnes (Mt) of CH_4 per year (2000 to 2009 annual means from IPCC AR5 WG1 Chapter 6, p. 507). Ranges of estimates assimilated by the IPCC are given in brackets. "% anthro. CH_4" is the percentage relative to total anthropogenic emissions.

Source	Mt CH_4 per yr	% anthro. CH_4
Fossil fuel production	96 [85–105]	29
Livestock	89 [87–94]	27
Landfills and waste	75 [67–90]	23
Rice cultivation	36 [33–40]	11
Biomass burning	35 [32–39]	11

From 2000 to 2009, 50–65% of global methane emissions came from human activities; the remaining 35–50% came from natural sources, mainly from anaerobically decomposing matter in wetlands. This source of methane may increase in the near future due to melting permafrost in northern regions, but the size of the increase is still highly uncertain.[36]

Table 3.1 gives methane emission estimates from human activities (annual means from 2000 to 2009). The largest anthropogenic sources of methane are fossil fuel production[37] (leakage) and livestock (75% of which is from cattle, who burp methane generated by fermentation in their digestive tracts). Each of these two sources accounts for between 4% and 10% of total anthropogenic greenhouse gas emissions, depending on choice of GWP time horizon.

Carbon dioxide, CO_2

CO_2 is the principal driver of global warming. Human emissions of CO_2 accounted for about half of the 2011 greenhouse gas radiative forcing (see Figure 3.7, page 50). When considered on the extended GWP_{100} basis instead of the instantaneous radiative forcing basis, however, CO_2 accounts for three-quarters of warming, because it remains in the atmosphere for a very long time.

Humans cause CO_2 emission by burning fossil fuels and re-moving forests. Approximately 90% of anthropogenic CO_2 currently[38] comes from burning fossil fuels, while approximately 10% comes from deforestation.[39] The amount of CO_2 released from removing forests is still uncertain, though, and current net CO_2 emissions from deforestation could range from 1% to 20% (with fossil fuel burning taking up the remainder). In the past, deforestation played a larger role. One-third of net *cumulative* emissions from 1750 to today are due to land-use change (mainly deforestation), which releases the carbon stored in wood and soils into the atmosphere via decomposition or fire.[40]

The residence of CO_2 in the atmosphere is complex, mediated by multiple processes transferring carbon between the reservoirs—the atmosphere, the ocean, the biosphere, and the rocks—on different time scales. (We'll discuss this "carbon cycle" in more detail below.) Because of this, the residence time of CO_2 isn't captured well by a single number. If humans stopped emitting CO_2 today, in a few hundred years, one-quarter or so of what we'd emitted would remain in the atmosphere, and in a few tens of thousands of years, one-tenth or so would remain.

Atmospheric CO_2 fraction has been precisely measured, to a small fraction of a part per million by volume (ppmv) high on Mauna Loa in Hawaii since 1958.[41] Today, the CO_2 fraction is measured at many different locations; as you'd expect, it tends to be a bit higher over areas of intense human activity. The annual highpoints in the Mauna Loa record are about two ppmv above the global average, Mauna Loa being in the northern hemisphere where 90% of humanity lives.

Atmospheric CO_2 fraction from Mauna Loa and from the global average are shown in Figure 3.3.[42] Notice the annual variation. The CO_2 fraction increases during the northern hemisphere winter months (from October to May) and decreases during the summer months (from May to October). Most of the world's plants are also in the northern hemisphere, and during the

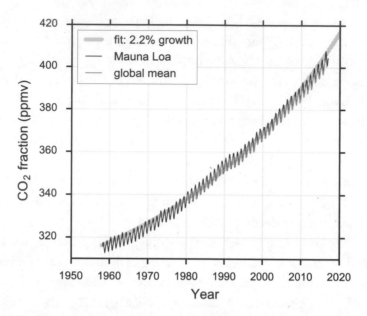

FIGURE 3.3. Atmospheric CO_2 fraction on Mauna Loa in Hawaii (black) and the global average CO_2 fraction (gray, traced over, and closely matching, the black curve), in the monthly average. The thick gray curve is the best exponential fit to the Mauna Loa data, with growth rate and start year allowed to vary. Best-fit growth rate is 2.2% per year, with best-fit starting year 1790.

northern summer months, plants are actively growing and incorporating CO_2 into their bodies. In the winter months, there is less growth but decomposition continues, releasing carbon back into the atmosphere via the tiny oxidative "fires" of biological respiration. Many have pointed out that this cyclical variation is like the biosphere breathing.

CO$_2$ fraction in our atmosphere has risen exponentially; the best exponential fit (thick gray curve in Figure 3.3) has growth starting in 1790, increasing annually at a rate of 2.2%.[43] In 1790, of course, James Watt had just succeeded in commercializing the steam engine. It's remarkable that an atmospheric CO_2 fraction record beginning in 1958 points back so precisely to humanity's fossil fuel revolution.

What about further back in time? Figure 3.4 shows the CO$_2$

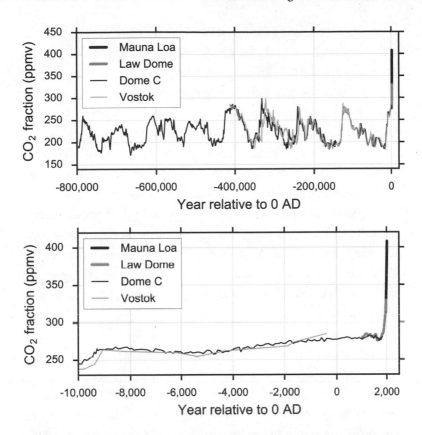

FIGURE 3.4. Ice core records combined with the Mauna Loa record; the bottom plot is a zoom of the right-hand edge of the top plot. The Mauna Loa record is shown in thick black, and only appears at the far right edge of the plots. The three ice core records are shown in black, gray, and thick gray (representing the Dome C core, the Vostok core, and the Law Dome core, respectively). Discrepancies between the black and gray curves give a sense of the uncertainty.

fraction going back to 800,000 years ago from three ice cores in Antarctica, including a zoomed-in view of the last 12,000 years. Ancient air bubbles trapped in the ice are analyzed for their CO_2 fraction, and time is inferred from the depth in the ice core.[44]

A few interesting things leap out of this record of nearly a million years of CO_2 fraction. First, CO_2 fraction is obviously much higher today than it has been at any time in the last 800,000 years. Second, the CO_2 fraction was remarkably constant over the last

11,000 years, allowing for the stable climate which supported the rise of agriculture and complex civilizations. Third, during this epoch, there were a dozen or so relatively rapid and significant rises in CO_2 fraction, although these were all far slower than the rise happening today, and stopped below 300 ppmv (whereas ours will clearly go far beyond 400 ppmv). For example, the rise beginning about 140,000 years ago took 10,000 years[45] from beginning to end, and maxed out at about 290 ppmv. These rises correspond to transitions from glacial periods to interglacial periods. Fourth, the CO_2 increases happened much more quickly than the CO_2 decreases, giving a sawtooth pattern to the record. Although reuptake of CO_2 is complicated and controlled by multiple processes, this suggests that our current warming might last tens of thousands of years. We'll see below that this is likely about right.

What humans do: Sources of global warming

What human activities cause the most greenhouse gas emissions, worldwide? Estimates for 2010 are shown in Table 3.2.[46] It's striking how human emissions come from a multitude of sources; no single line item dominates the list. This reflects the deep penetration of fossil fuels into today's predominantly industrial way of life.

The estimate for the agriculture, forestry, and other land-use (AFOLU) sector has relatively high uncertainty, as it is difficult to separate natural from human-made emissions. Livestock production (which includes clearing of land, manure decomposition, feed production, and enteric fermentation) accounts for 10–20% of humanity's global emissions.[47] The transportation sector is growing faster than any other, more than doubling between 1970 and 2010.[48]

Increasing global temperature

Figure 3.5 shows the average surface temperature of our planet since 1850, relative to the mean value of 1850–1900.[49] These data originate from land weather stations and ocean surface tempera-

TABLE 3.2. Mean anthropogenic global emissions by source, in 2010, as percentage of total. In 2010, we emitted a total of 49 gigatonnes (Gt) of CO_2e.

Source	Percentage of Total
Industry	**33%**
Electricity and heat	11%
Metals	5%
Chemicals	3%
Industrial waste	3%
Cement production	3%
Other industries	8%
AFOLU	**25%**
Land-use change and forestry	10%
Enteric fermentation	5%
Drained peat and peat fires	3%
Buildings	**18%**
Electricity and heat	12%
Transport	**14%**
Road	10%
Aviation	2%
Shipping	1%
Other	**10%**
Fuel production	6%
Petroleum refining	3%

ture estimates from ships and buoys, which are statistically combined to produce a global average estimate.[50] Care is taken to avoid biases from a variety of causes, such as imperfect sampling (you can't put a thermometer on every meter of the Earth's surface). However, despite this care they have been shown to be biased about 10% too low, if we interpret them as estimates of surface *air* temperature, as they use surface *water* temperature over the oceans.[51] Therefore, although this data set estimates that 2015 was 1.1°C to 1.2°C warmer than the historical mean, the truth was likely between 1.2°C and 1.3°C. Similarly, 2016 was likely 1.4°C to 1.5°C warmer than the historical mean.

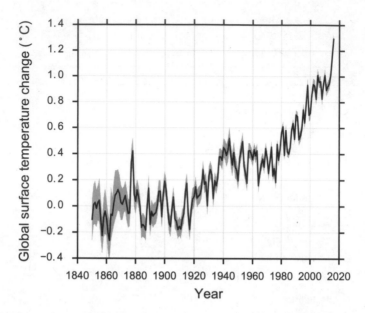

FIGURE 3.5. Global average surface temperature, relative to the 1850–1900 mean, through 2016. The shading gives a 90% confidence interval.

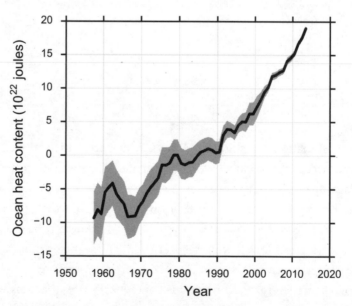

FIGURE 3.6. Global average ocean heat content, relative to the 1955–2006 mean. The shading gives a 90% confidence interval.

According to this data set, the 17 hottest years on record occurred between 1998 and 2016—a period of just 19 years.[52] This is remarkable given how noisy the data is. And there's no end in sight to this warming trend.[53] Today, people who simply point out this obvious trend in the data are often labeled "alarmist." In my opinion, it would be foolish *not* to be alarmed. So let me be perfectly clear: I'm a scientist, and I'm sounding an alarm.

Ocean heat content provides an even better measure of global warming, albeit less immediate to the human experience. The oceans absorb 93% of the heat pouring into the planet due to the energy imbalance from human greenhouse gas emissions, and ocean heat content has less variability than surface air temperature.[54] The global increase in ocean heat content since 1955, in the ocean layer from the surface down to a depth of 2,000 meters, is shown in Figure 3.6.[55]

Today's global warming is occurring more rapidly than any known warming in Earth's geologic history by more than a factor of ten.[56] Unsurprisingly, our planet is changing radically as a result. The rate of Greenland ice sheet loss has increased from 34 giga-tonnes (Gt) per year from 1992–2001 to 215 Gt per year from 2002–2011, a staggering six-fold increase in just a decade;[57] the ice sheet is lowering by 1.5 m per year around its edges.[58] Sea level has risen by about 20 cm and is rising at about ⅓ of a centimeter per year,[59] and accelerating.[60] The Arctic sea ice is roughly half gone (as of 2014, measured by summer minimum ice extent) and vanishing rapidly: summer ice extent is decreasing at a rate of between 9% and 14% per decade.[61] And, of course, hot days and heat waves are increasing in frequency and severity.[62] All of these changes are further evidence of warming, independent "thermometers." The Earth's warming is unequivocal, and the impacts are accelerating.[63]

But the climate system, of course, interacts complexly with the rest of the Earth system and is highly variable. Even as the global surface temperature increases, regional and temporal variations can create local or temporary colder weather. Indeed, changes in air and ocean circulation patterns caused by global warming could conceivably lead to regional cold anomalies. It's also quite possible for heat to be transferred from one part of the climate system to another, for example from the surface to ocean sublayers. It's important to remember that our ability to monitor the climate system is limited, and that particular variables (such as surface air temperature) impose their own idiosyncrasies and limitations on our view of the system.

The drivers of global warming

Radiative forcing estimates allow us to rank different drivers of global warming such as atmospheric CO_2, or changes in the sun's radiance. Figure 3.7 summarizes the state of our knowledge.[64]

It's worth carefully unpacking this figure, which gives radiative forcings in 2011 relative to 1750. First, note that 98% of the total change in forcing is human-caused. Most of this is due to human-emitted gases. The forcing from solar changes since 1750 are miniscule compared to human-caused changes; and over the last few decades, the sun has, in fact, been getting cooler.[65] In other words, relative to a few decades ago (as opposed to 1750), *more than* 100% of the forcing is attributable to humans.

Albedo changes due to land use (albedo means reflectivity) are from human-caused changes to the land surface, such as cutting down dark forests and replacing them with brighter crops and cities, which reflect more sunlight. These changes have a net cooling effect, and so are a negative radiative forcing. However, land-use change also causes CO_2 emissions, a positive forcing.

For each emitted compound, the figure includes indirect contributions from processes caused by the emission. For example, consider methane, CH_4. In addition to its direct greenhouse gas forcing, methane induces the creation of ozone (O_3) and stratospheric (high-altitude) water (H_2O^{str}), then eventually oxidizes to become CO_2. (As in the figure, I've listed these effects from strongest to weakest. Note that the shading in the figure's bars correspond to the "resulting drivers" as listed.) Ozone, water, and

CO_2 are also greenhouse gases. If we tallied these indirect effects somewhere else, we would underestimate the climate impact of human methane emissions. Halocarbons, on the other hand, destroy ozone, creating an ozone hole and producing a negative (cooling) forcing. Tallying this elsewhere would lead us to overestimate the effect of halocarbons.[66]

Note that human emissions of CO and NO_x don't contribute *directly* to the greenhouse gas forcing—they aren't greenhouse gases—but they do contribute *indirectly* via atmospheric chemistry (the creation and destruction of gases like CH_4, O_3, and CO_2).

Next, note that some aerosol pollution—especially sulfates and organic carbon from burning biomass—counteracts warming. These aerosol particles in the air reflect sunlight like tiny mirrors. I've referred to this group of aerosols as "bright aerosols" in the figure. They also act as cloud condensation nuclei, brightening clouds by inducing smaller but more numerous water droplets and increasing the cloud cover. These effects are "cloud changes due to aerosols."

You may have noticed the prominent plateau in warming between 1940 and 1978 in Figure 3.5; this is likely due to an increase in bright aerosol pollution.[67] Ironically, if we somehow suddenly eliminated these pollutants, the global warming forcing could conceivably jump up by almost 40%, although the uncertainty here is still quite large.

Black carbon, on the other hand, contributes significantly to warming. Black carbon is an aerosol produced in the agricultural burning of forests and savannas, residential biomass burning, and diesel engines. Instead of reflecting sunlight, black carbon absorbs it, giving a positive forcing of similar magnitude to methane's direct greenhouse gas effect.[68]

Finally, the figure gives the trend in total anthropogenic radiative forcing, from 1950 to 2011. We humans have nearly doubled our global warming impact since 1980. Radiative forcing is accelerating rapidly, driven mainly by exponentially increasing CO_2 emissions.

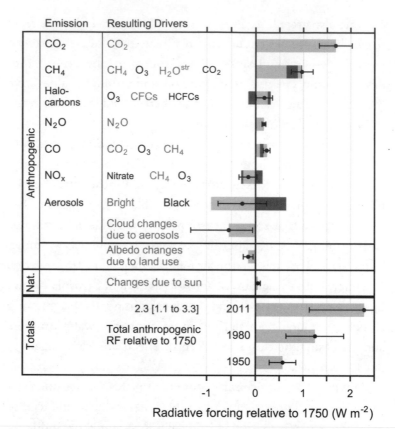

FIGURE 3.7. Radiative forcing estimates for the main drivers of global warming, in 2011 relative to 1750. The black dots indicate best estimates and the horizontal lines emanating from them indicate 90% confidence uncertainties.

Radiative forcings are half of the story. We're now ready to discuss the other half: how the Earth system changes as a result of these forcings.

Earth system feedbacks

Increasing greenhouse gas concentrations in the atmosphere force the Earth to a warmer state. Meanwhile, the Earth system changes in response to this forcing. The total amount of warming for a given greenhouse gas concentration—the climate sensitivity—depends on how the Earth responds to the forcing.

Processes that further enhance warming are called *positive feedbacks* (positive as in they go in the same direction as warming, not as in good), while processes that slow warming are called *negative feedbacks*. Together, forcings and feedbacks determine the temperature.

I've already mentioned a fundamental positive feedback: a warmer atmosphere holds more water vapor, which is a strong greenhouse gas. I've also mentioned a fundamental negative feedback: hotter objects lose more heat through infrared emission. Here are two other important feedbacks:

- **Surface albedo:** I've already mentioned albedo in the context of land-use change: when forests are destroyed and replaced with agriculture or human development, more sunlight is reflected to space (albedo increases). On the other hand, ice and snow reflect more sunlight than the ocean or soil exposed after it melts. This is a positive feedback: as ice and snow melt, the surface absorbs more sunlight, warming further. Albedo feedback due to sea ice loss is an important driver for changes occurring in the Arctic, which is warming more rapidly than the rest of the planet—and more rapidly than scientists expected.[69] Snow and ice albedo is also lowered by algae growth and buildup of anthropogenic soot.

- **Clouds:** Small changes in the global cloud cover can have large repercussions on climate. Clouds can block incoming sunlight (playing a huge role in global albedo) as well as outgoing infrared light; different kinds of clouds have different effects. Low-altitude clouds like cumulus (puffballs) and stratocumulus (overcast clouds) are warm and thick. They reflect sunlight, but because they're nearly as warm as the surface, they don't have a strong effect on the net upwelling infrared radiation; their net effect is cooling. Cold, wispy high-altitude clouds let sunlight through, but they absorb upwelling infrared light. Since these high cloud tops are cold, they don't emit much infrared. So their net effect is warming. The overall cloud feedback is likely positive: cloud changes are amplifying global warming.[70]

Surface albedo and clouds both contribute to Earth's global planetary albedo, which has a value of 0.29; Earth reflects a bit less than a third of the total incoming sunlight back into space. Global planetary albedo represents an average of all the ice, clouds, water, forests, cities, and deserts. Globally, clouds seem to act as a stabilizing buffer to the planetary albedo system in ways that we don't yet fully understand.[71]

There are additional climate feedbacks associated with the carbon cycle.

The Earth's carbon cycle

Carbon gets cycled through the atmosphere, the ocean, geological formations, and the bodies of all living things. Over the last four billion years of Earth's history, the processes of the carbon cycle have acted like a thermostat, tending to maintain a climate with liquid oceans, even when the sun was fainter in its youth. This stability has allowed Earth's biodiversity to flourish.

> People ask me if they should be especially worried about methane. I tell them there's plenty to worry about from human-released CO_2.
>
> — DAVID ARCHER, climate scientist

On shorter time scales, however, the carbon cycle amplifies abrupt climate changes; these climate changes played an important role in the biosphere's previous mass extinctions.[72] These past examples of carbon cycle amplification inform our current global warming event.

There are several major reservoirs of carbon on Earth that vary widely in size and interact with each other and with the climate system on vastly different time scales; see Table 3.3.[73] Before the industrial era, carbon flows between these reservoirs were roughly in balance—no net flows. Now, because of human emissions, there are imbalances and net flows. The CO_2 we emit accumulates in the atmosphere, but also spills into other reservoirs on various time scales. Currently about 57% of the CO_2 we emit does not end up in the atmosphere;[74] instead, it's absorbed by the ocean, biomass, and soils, all of which are acting as "carbon sinks." In general, these

TABLE 3.3. Carbon reservoirs.

Reservoir	Size (GtC)	Note
Atmosphere (1750)	589	
Atmosphere (2011)	830	increasing at 4 GtC/yr in 2011
Ocean waters	38,000	
Ocean sediments	2000	calcium carbonate
Ocean methane	1500–7000	frozen in deep sediments
Biomass	450–650	mostly trees
Soils	1500–2400	
Fossil fuels	~13,000	resources plus reserves
Coal reserves	450–540	
Gas reserves	380—1100	
Oil reserves	170—260	
Permafrost	2000	frozen but thawing peat
Limestones	>60,000,000	
Kerogens	15,000,000	fossil fuel precursors

carbon sinks are poorly understood; but it seems likely that over the next few decades or centuries they'll decrease in effectiveness, as the ocean becomes increasingly saturated and the land reservoirs respond to a warmer world.[75]

Here's a brief picture of the carbon cycle in terms of its reservoirs:

- **The atmosphere:** While small in size, this key reservoir interconnects many parts of the carbon system, a Grand Central Station of the carbon cycle. Carbon flows in from the ocean, the respiring biosphere, and from burning fossil fuels and forests; and carbon flows out into the ocean and into the growing biosphere (mainly trees). Because of additional emissions from humans, in the 2000s carbon was accumulating in the atmosphere at an average rate of 4.0±0.2 GtC (gigatonnes of carbon) per year.[76] This accumulation rate is likely to increase due to our accelerating emissions, and also due to a slowdown in uptake from the ocean and land reservoirs.
- **The ocean:** About 28% of the total CO_2 we've emitted so far has been absorbed by the ocean and converted into carbonic

acid.[77] As the ocean warms and becomes increasingly saturated with CO_2 it will become a less efficient sink. This means more of our emitted CO_2 will remain in the atmosphere, accelerating warming. There is some evidence that the ocean carbon sink is already becoming saturated,[78] but our understanding of how the ocean sink works and how it will work in the future remains uncertain.

Indeed, we know (from isotopic analysis) that the ice age cycle was driven by a positive feedback loop involving ocean uptake of CO_2. During glacial times, the colder ocean somehow pulled more CO_2 out of the atmosphere than can be explained by the temperature dependence described above. However, we don't know what additional process (or processes) were operating. We can't eliminate the possibility that this additional unknown feedback might also operate in reverse, contributing to warming.[79]

- **The land:** Soil and plants (which are intimately interconnected), together with permafrost, make up the land carbon reservoir. Increased atmospheric CO_2 augments plant growth and therefore plant carbon uptake, up to a point.[80] We don't know how to directly measure the size of this land carbon sink, but we infer that, on average, about 29% of the total CO_2 we've emitted has been absorbed by the land sink (that is, the total sink, 57%, minus the ocean sink, 28%). Annual variability of land carbon uptake is high. During some El Nino years, the land "sink" ends up being a net source; in other years, it absorbs more than 29%. It's still unknown where and how this absorbed carbon is apportioned between tropical forests and boreal forests. Unfortunately, however, there's strong empirical evidence that future warming will drive a net loss of soil carbon to the atmosphere,[81] and that global forest mortality and wildfires will increase in the future—and that these sources of atmospheric carbon have been underrepresented in projections to date.[82] We might see the catastrophic loss of the Amazon rain forest and release of its carbon into the atmosphere within decades, as drought makes the forest increasingly susceptible to fire,[83] and projected severe drying in the 21st century pushes it toward a tipping point.[84]

About a quarter of the northern hemisphere is covered in permafrost. As the permafrost thaws, microbes convert the biomass (mostly peat) into CO_2 if oxygen is present, and CH_4 if oxygen isn't present. It's estimated that there are about 1,700 GtC in the permafrost, more than four times what we've emitted from burning fossil fuels. The rates of permafrost CO_2 and methane release as the planet warms are unknown.

It's likely that the global land carbon sink will switch from being a net carbon sink to a net carbon source, accelerating climate change. However, as with our understanding of the ocean sink's future, uncertainty about the details is high.[85]

- **Frozen ocean methane:** Methane is frozen in ocean sediments on continental shelves. There's a lot of it, thousands of GtC worth, many times the carbon that humanity has so far emitted. If the Earth warmed enough to begin releasing this methane, it would be a positive feedback. Everything about ocean methane release is uncertain—mechanism, magnitude, and timing—but we might be wise to plan for at least 0.5°C additional warming from ocean methane release after 3°C of global warming, and this warming boost could last for thousands of years.[86]

Most climate models don't yet include the carbon-thawing feedbacks (permafrost and frozen ocean methane), biasing their predictions of future warming toward being too conservative. Slow release from these reservoirs could ultimately end up contributing about the same amount of CO_2 to the atmosphere as humans end up emitting, effectively doubling our emissions; exactly how this plays out, though, is anyone's guess.[87]

- **Limestones:** Carbonate rock (limestone) is the largest carbon reservoir, and it interacts with the climate system on tectonic time scales, as we'll discuss below.

- **Fossil fuels:** Humanity had burned a total of 375±30 GtC of fossil fuels as of 2011, and in one year (2011), we burned 9.2 GtC.[88] Between 2000 and 2010, our global greenhouse gas emissions accelerated exponentially at 2.2% per year;[89] this increase was driven by an increase in our burn rate. (Recall that the measured atmospheric CO_2 fraction has this same growth rate over the long term; see Figure 3.3, page 42.)
- **Kerogens:** When organisms die on land or in the ocean, their carbon can become trapped in kerogens—fossil fuel precursors—which can eventually transform into coal, oil, and natural gas. As with the rock sink processes, these processes operate on long time scales. Both the rock and the kerogen sinks will be irrelevant on human time scales, but over millions of years, they will be significant. Perhaps in ten or 100 million years, Earth's "battery" will once again be charged, with much of the fossil carbon that humans dug up and burned again buried underground in the form of coal, oil, and gas.

Carbon flows

How does carbon flow between these reservoirs? Table 3.4 lists the flows of carbon in and out of just one of the reservoirs: the atmosphere.[90] This table separates these flows into estimates of human-caused and natural (i.e., preindustrial) portions. The much larger human-caused flows dramatically demonstrate the dominance of modern human emissions over the carbon cycle.

The amount of atmospheric carbon is a direct driver of climate change. Over time scales of thousands of years and more, atmospheric exchange of carbon with the other reservoirs determines this amount, as the carbon cycle recovers a dynamic equilibrium.

One such slow flow is the weathering of rocks on land. Rain is weakly acidic (it contains dissolved CO_2); it slowly dissolves exposed carbonate and silicate rocks, creating calcium ions and bicarbonate ions which eventually wash into the oceans. Marine organisms use these materials to build their shells, which then accumulate on the ocean floor, forming limestones. On geologic

TABLE 3.4. Carbon fluxes in and out of the atmosphere in GtC per year. Human-caused fluxes are the annual mean from 2002 to 2011.

Source	Flux (GtC/yr)
Human-caused fluxes	
Net atmospheric increase	4.3 ± 0.2
Fossil fuel burn (2011)	8.3 ± 0.7
Ocean-to-atmosphere (total)	−2.4 ± 0.7
Land-to-atmosphere (total)	−1.6 ± 1.0
Net land-use change	0.9 ± 0.8
Residual land sink	−2.5 ± 1.3
Additional natural fluxes	
Volcanism	0.1
Weathering	−0.3
Ocean-to-atmosphere	0.7
Land-to-atmosphere	−1.7
Freshwater outgassing	1.0

time scales, the limestones are subducted under the Earth's crust. Subjected to heat and pressure, they morph back into silicates, releasing CO_2 which cycles back into the atmosphere via volcanic venting.

This weathering cycle is an important part of the Earth's thermostat system. In the distant past, almost a billion years ago, there may have been a period when the planet was a frozen "snowball Earth." Volcanic emission of CO_2 and reduced rock weathering are thought to have helped bring the Earth out of this period. This flow is significant on time scales of hundreds of thousands of years.

Another flow, not as slow, involves the ocean floor sediments which contain accumulated calcium carbonate shells. CO_2 dissolves in the ocean by combining with carbonate ions to make carbonic acid, which dissociates into bicarbonate.[91] This makes the water more acidic, which then dissolves the sediments,[92] releasing more carbonate ions which are then available to react with CO_2. The ocean floor sediment buffer therefore allows the ocean to dissolve more CO_2 than it otherwise would. There's a catch though: we'll need to wait thousands of years for this exchange to help us, as the time scale for the acidic surface waters to reach the sediments is set by the time scale for deep ocean water mixing. We'll discuss the lingering fate of our CO_2 emissions in the next chapter.

Summary: The cause of global warming

Multiple lines of evidence show that humans are causing the observed exponential increase in atmospheric CO_2. For example, we know roughly how much CO_2 we emit each year from burning fossil fuels because the fossil fuel industry keeps track.[93] We also know that about one-third of our total emissions have been from deforestation. We can thus directly compare an estimate of our accumulating CO_2 emissions to what we see in the atmosphere.[94] To account for deforestation, I've multiplied the fossil fuel data by a factor of 1.3 and plotted this as a dotted black line in Figure 3.8, alongside the actual observations from Mauna Loa (gray line; we discussed these observations above).[95]

That dotted black line shows that since 1965 humans have emitted about twice the CO_2 that remains in the atmosphere. As discussed earlier, we know that 28% goes into the ocean, and the rest (29%) must be going into the land because there's no other plausible place for it to go. The solid black line accounts for these carbon sinks, in good agreement with observations. This provides unequivocal evidence that humans are the cause of the observed CO_2 increase.

In case that's not enough, here are four additional lines of evidence: 1) decreasing atmospheric oxygen is consistent with what we'd expect from a fossil fuel source; 2) the downward trend over time of the $^{13}C/^{12}C$ isotopic ratio is consistent with what we'd expect from a fossil fuel source; 3) the downward trend over time of the $^{14}C/^{12}C$ radiocarbon isotopic ratio is consistent with what we'd expect from fossil fuel and nuclear testing sources; 4) the difference in atmospheric CO_2 concentration measured in the northern and southern hemispheres, and the increasing trend in this difference over time, are both consistent with what we'd expect from a fossil fuel source.[96]

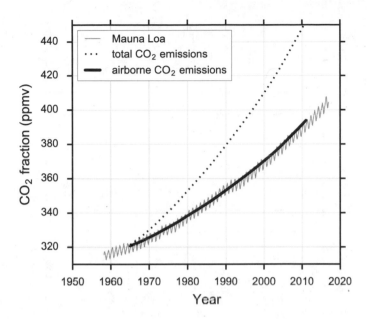

FIGURE 3.8. Observed atmospheric CO_2 fraction as measured on Mauna Loa (gray line), the calculated atmospheric fraction if 100% of CO_2 emitted by humans since 1965 stayed in the atmosphere (black dotted line), and the calculated atmospheric fraction accounting for a net 43% carbon sink (black line). This figure demonstrates the correspondence between our CO_2 emissions and the CO_2 that we measure in the atmosphere.

How can we be certain that this human-emitted CO_2 is indeed the primary cause of global warming? Because of the complexity of the Earth system, and because there are other radiative forcings, it's not simple to predict the amount of temperature increase for a given CO_2 increase. So this is a more difficult question.

Personally, I'm certain that the CO_2 is causing warming, and so are 100% of my immediate colleagues. Here's why. First, a clear and strong causal link is contained in the basic physics of the greenhouse effect. We understand how CO_2 interacts with infrared radiation extremely well, and the observed global warming is consistent with basic physics. An absence of warming, on the other hand, would be surprising and require explanation. Interestingly, this basic science has been known for over 120 years.[97]

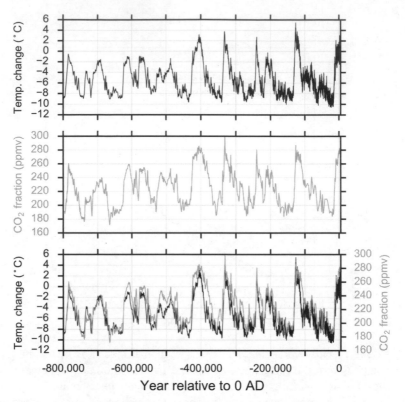

FIGURE 3.9. Antarctic surface temperature change and atmospheric CO_2 fraction from the Dome C Antarctic ice core. The temperature change is relative to the average over the last 1,000 years. The most recent sample in the record is dated to 1911; these plots therefore don't include the modern spikes in both temperature and CO_2 fraction.

A second line of evidence comes from the ice core samples. We've already seen that, by analyzing trapped bubbles of ancient atmosphere, we can directly measure the atmospheric CO_2 fraction back to 800,000 years ago; we can also infer the Antarctic surface temperature record from water isotopes.[98] Figure 3.9[99] shows a strikingly clear correlation between temperature and atmospheric CO_2 fraction. Correlation doesn't imply causation, and in fact the ice core CO_2 changes lag the temperature changes and are thought to be amplifiers instead of initial drivers;[100] but the correlation certainly does mean that a present-day lack of warming, given the observed human-caused CO_2 spike, would require explanation. CO_2 *initiating* warming might be unique to today's ongoing climate change.[101]

The third link comes from models. As we've seen, there are many factors controlling the global temperature in addition to the CO_2 fraction. For one thing, other human-emitted greenhouse gases add to the warming, and human-emitted aerosol pollutants reflect sunlight and block some of the warming. There are clouds and other Earth system feedbacks, there are multiple carbon sinks and natural carbon sources, there are small changes in the sun's intensity, and there are even volcanoes sporadically emitting cooling aerosols. Earth system models attempt to account for as many of these complex and interacting processes as they can. Of course, their predictions differ, as we explicitly saw in Figure 3.2, page 33, but every Earth system model says that increasing the CO_2 fraction leads to warming.

In fact, no Earth system model is able to replicate the observed warming in the absence of rising CO_2, and there's no alternative explanation for the warming that's consistent with observations. It's my opinion, therefore, that any reasonable person must accept that human-emitted CO_2 is a significant cause of the observed global warming—unless equally strong evidence to the contrary comes in. So far, I know of no such evidence.

CHAPTER 4

Global Warming: The Outlook

*There's already enough science
to know what we need to do,
and we're not doing it.*

— KEN CALDEIRA, climate scientist

In Chapter 3, I described how we know that human-emitted CO_2 is the primary driver for the observed global warming trend. In this chapter, I'll briefly describe some of the ramifications of this warming for humans and nonhumans. As before, I won't attempt to be complete.[1] My goals in doing so are to further describe what we know and what we don't know, and to further clarify the likely duration of global warming.

Impacts

Global warming is bringing some benefits. Farmers and gardeners in cold regions have longer growing seasons and warmer soils; corporations will soon be able ship their products across an ice-free Arctic Ocean; death rates due to cold weather will likely decline.

In my opinion, though, the harmful impacts of global warming outweigh these benefits, and they will be permanent for all human purposes (as we'll discuss below). If we decide we don't like a warmer world, there will be no going back. The precautionary principle applies here, in spades.

Most of the impacts aren't all-or-nothing. For most, costs will mount with temperature. Every degree of global warming will make human life on Earth more difficult.

Heat waves

The number of hot days and nights in a year have increased, and heat waves are more frequent and last longer.[2] Europe's 2003 heat wave, which killed 70,000 people,[3] was the first individual event statistically attributed to global warming.[4] In 2014, heat waves were already ten times as likely as they were just ten years earlier, and this frequency is increasing.[5]

Sixth mass extinction

Global warming, along with the sheer space taken up by so many humans, is driving an ongoing mass extinction event. Plants and animals are migrating to cooler habitats; those that can't move, or move quickly enough, are vanishing. There are numerous scientific studies detailing the rapid migration of animals and plants poleward at an average rate of over a mile per year, and where possible to higher elevations.[6] Marine species are also moving poleward, and doing so even more rapidly than land species.[7]

It has been estimated that, by 2050, global warming alone, depending on how aggressively we try to stop it, will commit 15% to 37% of all species to extinction.[8] Interestingly, though, the fossil record from the time of the Paleocene-Eocene Thermal Maximum (PETM) global warming event does not show a corresponding mass extinction.[9] This might be because warming happened 100 times more slowly in the PETM, giving species time to adapt.

In his book *Half-Earth*, biologist E. O. Wilson wrote, "Biodiversity as a whole forms a shield protecting each of the species that together compose it, ourselves included.... As more and more species vanish or drop to near extinction, the rate of extinction of the survivors accelerates."[10] Today's extinction rate is already about 1,000 times the background rate, and the extinction pressure from global warming is accelerating.[11]

Ocean acidification

Because a quarter of our CO_2 emissions dissolve into the ocean and form carbonic acid, the ocean is becoming more acidic. Indeed, global mean ocean pH departed from its historical bounds in 2008; ocean dwellers are already living in a new world.[12] Acidification and rising water temperatures have devastated the world's coral reefs, a blow to global biodiversity. Other ocean ecosystems are also changing.

Destructive weather

Global warming is changing the hydrologic cycle. A warmer atmosphere holds more water, and the amount of precipitation over northern hemisphere land masses has increased.[13] In some regions, including across the US, this increased precipitation has translated into a higher chance of destructive rains;[14] parts of the US could experience a five-fold increase in the probability of extreme storms, which will also become more intense.[15] Regional analyses are finding a decreasing trend in the wintertime snow-to-rain ratio.[16]

Global warming is also changing large-scale atmospheric circulation patterns,[17] leading to increased snow and rain in some regions, increased risk of drought in other regions,[18] and changes in monsoon patterns and monsoon weakening.[19] It seems that wet areas are getting wetter and are more prone to flooding from heavy storms, while dry areas are getting drier.

Drought, fires, and food

Hotter temperatures dry out the soil more quickly, and when combined with lack of precipitation, produce drought. Drought is causing unprecedented wildfire in many regions of the world. In particular, drought is projected to worsen in the Southwest and Central Plains of the US over the coming decades.[20] Due to a multiyear drought, 2015 was the worst fire year on record in the US. As of 2016, over 100 million trees had died due to California's worst drought in over 1,200 years, which was driven by record temperatures.[21] This unprecedented and worsening forest die-off might signal a permanent climate-driven ecologic shift.[22]

Climate change is now the primary driver for forest dryness and the longer fire season in the Western US, and has doubled the area that otherwise would have burned since 1984.[23] Wildfire area increases exponentially with global warming, with each degree of warming leading to more fire than the last, at least while there are still trees left to burn.[24]

Drought also affects crop yields, as do heat waves. Although warming brings some benefits to high latitude crops, negative impacts on crop yields outweigh positive impacts globally, and warming is already reducing global wheat and maize yields.[25]

When temperature and precipitation regimes both change, there's little in agriculture that *isn't* affected. Here's one example that might wake you up: by 2050, the area suitable for coffee production could be half as large as it was in 2016.[26]

Sea level

Coastal and low-lying areas have an increasing risk of flooding and submergence as sea levels continue to rise due to ice loss and thermal expansion.[27] Sea level rise will not be globally uniform; some regions will experience more rise than others.[28] Areas most at risk, in terms of overall cost of damage, include Miami, New York, New Orleans, Boston, Mumbai, and Guangzhou.[29] These and other cities will eventually be forced to build costly protective structures, or risk abandonment; this could swamp the real estate market.[30] Without adequate adaptation, hundreds of millions of people will be displaced by 2100.[31]

As this gradual sea level rise continues to accelerate, we also face additional, unpredictable increases due to ice sheet disintegration. For example, collapse of the West Antarctic ice sheet, which may now be inevitable due to past and ongoing destabilization, will raise sea level by at least 3 meters.[32] But we don't yet understand how the collapse of this and other ice sheets might unfold. Some models predict that West Antarctic collapse will take centuries.[33] Another predicts decades, and the model underlying this faster prediction skillfully reproduces sea level rise in the geologic record.[34] In the coming years, I'm certain that we'll learn a great deal about ice sheet collapse, from both models and observations.

Human migrations and civil unrest

Global warming exacerbated a serious drought in Syria which helped precipitate civil war.[35] While it's important to keep in mind that warfare always arises from a mixture of factors, there's nonetheless a clear link between climate change and increased risk of warfare,[36] a risk that the Pentagon is taking very seriously.[37]

Pressures from sea level rise, drought, and famine due to global warming increase the risk of mass migration. Future climate-related migrations, for example from low-lying Bangladesh, could dwarf earlier refugee crises.

Socioeconomic distribution of impacts

How these impacts affect humans depends not only on warming, but on social structures. Financially poor and marginalized people are more vulnerable, and financially poor countries are less able to adapt and respond to climate crises.[38] This fact, together with the fact that rich countries have emitted far more greenhouse gases *per capita*, adds a deep ethical dimension to our choice of response to global warming. This isn't to say the rich are safe from these impacts. After enough warming, everyone will be vulnerable.

How long will global warming last?

The duration of global warming depends on how much CO_2 we, choose to emit, and how our emissions interact with the carbon cycle over time. Imagine we emit a pulse of CO_2 into the atmosphere (it isn't hard to do). After a few years, the CO_2 would become well-mixed in the atmosphere; and several processes in the carbon cycle would pull it from the atmosphere and into other carbon reservoirs on a variety of time scales: plant and soil uptake, ocean uptake, ocean floor sediment dissolution, and rock weathering.

As we saw in Chapter 3, plants and the ocean will remove roughly half of the emitted CO_2 on a time scale of a few decades. What the plant and soil system will do on a longer time scale (say, a century) is a topic of active research: there is a good chance it will release some of the absorbed CO_2 back into the atmosphere through fires and permafrost thawing.

The ocean will continue to absorb the CO_2 on this century time scale. The precise fraction it absorbs depends on how much we emit. The more we emit, the higher the fraction of CO_2 that will stay in the air (a warmer ocean holds less). The new equilibrium airborne fraction, achieved after a couple of centuries, will likely be between 15% and 40% and persist at this level for a few millennia. If we mitigate aggressively, it will be closer to 15%. If not, it will be closer to 40%.[39]

After this initial stage of rapid uptake by the ocean ends, the slower processes will take over. For the next few tens of thousands of years, calcite sediments on the seafloor will gradually dissolve in the acidified ocean, allowing it to draw more CO_2 from the atmosphere until 10% or so of the CO_2 remains. Then, for the next few hundreds of thousands of years, calcium carbonate precipitation in the ocean, driven by rock weathering on land, will finish the job.[40]

How can we predict that the ultimate time scale for CO_2 recovery is a few hundred thousand years? Aside from our models, which continue to improve, we have an extraordinary historical event with which to test them. The PETM was an abrupt global warming event that occurred about 56 million years ago. Its signature is clearly visible in ocean sediment cores.

Over the course of 8,000 years after the PETM, the Earth's temperature rose by about 5°C to 8°C due to elevated CO_2 in the atmosphere. (What triggered this elevated CO_2 in the first place is still a mystery.) The CO_2 fraction gradually recovered over the next 200,000 years.[41]

The PETM is the historical event most nearly matching today's human-driven event. It's the fastest climate change event we know of, except for today's event—which is happening 10 to 100 times as rapidly, amplifying stresses to species and ecosystems.

Ten million years

It will take fifty or a hundred times this length of time, however, for biodiversity to recover. We know from fossil records that the biosphere takes ten million years to recover its biodiversity after mass extinctions.[42] These two time scales—100,000 years for CO_2 reuptake and 10 million years for biodiversity recovery—aren't long from the Earth's point of view. But they're both good approximations to "forever" from the perspective of human lives or even human civilizations.

One of my main purposes in writing this book is to suggest that a good overarching goal for today's civilization would be to minimize global warming and its concomitant biodiversity loss for the sake of the next few hundred thousand human generations.

At first, we didn't know what we were doing. It was reasonable for us to start burning fossil fuels. However, now we do know what we're doing. It bothers me to burn fossil fuels, and it also bothers most of my colleagues in Earth science.[43] I've personally concluded that continuing to mindlessly burn fossil fuels, knowing what I know, is unacceptable.

Geoengineering

Given the reality of global warming, those of us who judge it to be undesirable might reasonably hope that humanity will find a quick technological fix. This hope resonates with deep aspects of our modern industrial mindset, such as the fetishization of technology, the supremacy of convenience, and the desire to control nature. However, I currently see no evidence that geoengineering will save us; and in my opinion, it's a mistake to hope that it will.

> We can't solve problems by using the same kind of thinking that created them.
>
> — attributed to ALBERT EINSTEIN

There are two broad categories of geoengineering ideas: those aiming to reflect sunlight and those aiming to take CO_2 out of the atmosphere (carbon capture).

The leading carbon capture geoengineering technique,[44] and the best idea I've seen so far, involves increasing the rate of weathering of silicate rocks by quarrying them, pulverizing them, and spreading them on fields in the tropics, where they would slowly absorb atmospheric CO_2.[45] But the scale of this operation would be massive, many times larger than the current scale of fossil fuel extraction.[46] It would be extraordinarily expensive, require huge amounts of energy (either releasing yet more CO_2 or requiring significantly more renewable energy infrastructure), and it could adversely affect soils. Furthermore, we still don't have a good understanding of weathering's role in the carbon cycle: it's unclear, for example, to what extent the long-term climate is controlled by the weathering of silicate rocks as opposed to carbonate rocks.[47]

It doesn't make sense to expend huge amounts of energy and effort to take CO_2 out of the atmosphere on the one hand, while we're still releasing it by burning fossil fuels at an accelerating pace on the other. Perhaps someday we will implement large-scale renewable-energy-driven silicate weathering—after we've decarbonized our energy system.

The sunlight reflection approach could be implemented much more quickly and cheaply, but it has serious drawbacks. Members of the Royal Society (and others) consider airplane dispersal of sulfate aerosols into the stratosphere to be the "nearest approximation" to an "ideal method" of geoengineering.[48] While this seems likely to lower global surface temperatures—we can't be sure, because we don't yet have a deep understanding of the albedo system, and our naïve attempts to engineer albedo might be buffered by compensating processes[49]—it would also carry great risks. No one knows all of the side effects it would impose on the Earth system, but global models suggest that it would lead to changes in the precipitation cycle, adversely affecting rainfall in some regions.[50]

This would create issues of international fairness. Certainly, wars have been fought for less; and the stakes of such water wars might be especially high, given the Indian subcontinent's dependence on monsoon rains and the nuclear tensions there. Further, it's fundamentally impossible for us to know all of the side effects in advance with models. The aerosol geoengineering option would amount to closing our eyes and jumping.

Another concern that I have, personally, is that we humans would collectively take the artificially lowered surface temperature as an excuse to continue postponing meaningful mitigation efforts. This combination of suppressed surface temperatures and lack of mitigation would leave the biosphere vulnerable to a sudden temperature spike due to an unforeseen discontinuation of the geoengineering effort. For example, given the confluence of factors we'll discuss in Chapter 5, it might be unwise to rule out the possibility of significant economic and political collapse. In the event of such a collapse, would the geoengineering planes keep flying? Steadily accelerating atmospheric accumulations of both CO_2 and sulfate aerosol pollutants with the looming prospect of sudden cessation—what could go wrong?

Geoengineering prospects go downhill from here. Most other ideas, such as "space sunshades," seem scarcely realistic or technologically realizable. For example, one proposal calls for "a swarm of trillions of thin metallic reflecting disks each about 50 cm in diameter, fabricated in space from near-Earth asteroids."[51]

A better and easier idea, in my opinion, is to *stop burning fossil fuels*. It's possible, as we'll see in the second part of this book. The sooner we stop burning fossil fuels, the better off we'll be.

Agricultural soil carbon sequestration

In Chapter 3, we discussed early experimental evidence that warming will likely drive a net release of CO_2 from global soils, mainly from Arctic and subarctic regions, although the magnitude of this release is still very uncertain. Keeping that in mind, what are the prospects for using soils to pull CO_2 out of the atmosphere by adopting agricultural methods such as no-till, cover cropping, manuring, biochar, and improved grazing?

No one really knows. One study estimates uptake would be modest, perhaps offsetting 5% to 15% of global emissions annually;[52] a more recent study puts the estimate at between 0% and 6%.[53] This uptake would taper off until the soil reached a new equilibrium after between 20 and 100 years; the agricultural practices would then need to continue indefinitely to prevent rerelease back to the atmosphere.

Further research is needed in this area, but it would be a mistake to expect "carbon farming" to be a silver bullet. The simple fact remains: we must stop burning fossil fuels.

How much more fossil fuel can we burn?

Scientists can imagine various mitigation pathways—particular mixes of technology, conservation, and human decisions—and then model emissions and temperature trajectories for these pathways.

The IPCC AR5 WG3 considered about 900 such pathways and translated them into actionable milestones. For example, to be on track for the RCP 4.5 scenario, with an estimated 66% chance at keeping peak warming under 3°C, at least 40% of humanity's total energy supply must be carbon-free by 2050. To be on track for the RCP 2.6 scenario, with a 66% chance at keeping peak warming under 2°C, at least 60% of humanity's total energy must be carbon-free by 2050.[54]

Unfortunately, the scenarios that would give us a fighting chance of remaining below 2°C aren't realistic. Out of the 400 such scenarios, 344 assume large-scale deployment of negative-emission technologies such as carbon sequestration and rock-weathering geoengineering.[55] The problem is that such technologies are speculative; they don't yet exist. Also, each of the remaining 56 scenarios relies on global emissions peaking in 2010—but emissions didn't peak in 2010, so these scenarios are already invalid. This kind of misplaced optimism does a disservice to humanity, because it lulls policymakers and the public into a false sense of security.

How much more fossil fuel, then, can we burn and maintain a decent chance of keeping warming below 2°C? For a 66% chance of staying under 2°C, we can emit a total of 270 [175 to 338] GtC from 2011 onward.[56] This includes everything: burning fossil fuels, burning forests, and making cement.

But emissions between 2002 and 2012 were 9.2 GtC per year on average, and growing exponentially at 2.2% per year. In just the five years from 2011 to 2016, therefore, we emitted at least another 46 GtC. Let's optimistically assume that sometime between 2016 and 2100 we begin regrowing forests, and we do so well that there's no net deforestation in this period; and that we constrain our cement emissions to 40 GtC from now until we phase out cement by 2100.[57] This leaves 180 [89 to 252] GtC for burning fossil fuel. That's our fossil fuel budget from 2016.

In 2010, we emitted 8.7 GtC from burning fossil fuels. At the 2.2% growth rate, we'll burn through our budget by 2031 [2024 to 2036]. How quickly do we need to ramp down our emissions to stay within budget?

Suppose we'd started in 2016. We'd have needed to ramp down by 6% per year (or 12% per year in the worst-case scenario, where the 2016 budget turned out to be only 89 GtC). And the longer we wait, the more quickly we'll need to ramp down. If we wait until 2020, we'd then need to reduce by at least 10% per year (or 40% per year in the worst-case scenario).

Any way you slice it, keeping warming below 2°C requires an immediate, massive, and global mitigation effort. With each passing day, it's less likely that we'll succeed. Indeed, over 90% of Earth scientists believe we'll surpass this threshold.[58]

Clearly, production of unconventional resources such as the Alberta tar sands or Arctic oil is not consistent with the 2°C goal. If we need to leave huge fractions of our fossil fuel reserves in the

We need to leave huge fractions
of our fossil fuel reserves in the ground.

ground, the rational response would be to immediately give up exploration and development of these less economic resources, which we won't need if we're indeed planning to ramp down.[59] That we're in fact pursuing them is not cause for optimism.

I hope this discussion has given a clear and informed sense of urgency.

Causes for optimism

Despite this urgency, I feel optimistic about meaningful collective action on several fronts. One is renewable electricity, which is growing rapidly. One hundred percent carbon-free electricity is within reach in the US; I estimate that we could make our entire electrical grid carbon-free for a few thousand dollars per person, far less than we've already spent on the war on terror.[60] Of course, US electricity is only part of a much bigger picture which includes food, forestry, development, manufacturing, transportation—and the world outside of the US. But by 2016, one-quarter of global electricity was generated renewably.[61]

Another is California, my home state and arguably the world's 6th-largest economy, which is on track to be 40% below 1990 emissions levels by 2030.[62] California is actually doing something: setting meaningful goals and achieving them. Other states and countries should do the same.

A third is the prospect of the US someday adopting a national revenue-neutral carbon fee (see Chapter 14), which would greatly accelerate the transition to a carbon-free energy system and cause this transition to pervade every aspect of our lives. When the US adopts a revenue-neutral carbon fee, other nations would likely follow suit.

Uncertainty

In this and the previous chapter, I've attempted to point out which scientific details are certain and which are still uncertain. However, it's a mistake to think that successful collective action depends on reducing the scientific uncertainty surrounding some of the details of global warming. For one thing, the overall scientific picture is unequivocal[63]—continuing on our current path will lead to catastrophic changes to the Earth system that gives us life. It is this overall picture which demands action.

The remaining scientific uncertainties are now overshadowed by cultural, economic, and political uncertainties. Indeed, the largest uncertainty of all is how much fossil fuels humans will choose to burn.

We climate scientists have done good work. We've concluded, unequivocally, that global warming is caused by humans burning fossil fuels, although we continue to understate the urgency and seriousness of the threat by systematically "erring on the side of least drama."[64] Although we'd do well to learn how to connect more effectively with the public—to become storytellers—we have nonetheless delivered our message. How humanity acts on that message is up to the policymakers, which really just means it's up to people like you.

CHAPTER 5

Growth Always Ends

*The greatest shortcoming of the human race is
our inability to understand the exponential function.*

— ALBERT ALLEN BARTLETT

Sooner or later, anything that physically grows must stop. This is
true of plants, animals, and colonies of bacteria. It's equally true of
civilizations. And if the rate of growth is exponential, the end can
come with surprising rapidity.

In this chapter, we'll discuss how we misperceive exponential
growth, and how human growth is finally bumping up against the
hard physical limits of our biosphere. One way or another, and
sooner or later, these limits will drive a radical transformation of
human civilization.

Many people assume that the human population can grow exponentially forever because we'll soon populate planets orbiting
other stars, or that our economy can grow exponentially forever
by decoupling from physical resources. Not only are these techno-
dreams unrealistic,[1] they're dangerous distractions. Instead, I hold
a different dream: living well on Earth without growth.

Explosive exponentials

Every day, something growing linearly adds a fixed amount, but
something growing exponentially *multiplies by* a fixed amount. I've
noticed that my brain extrapolates growth by assuming trends will
follow a straight line over time. For most processes, this is a good

short-term approximation. But if this is true of most human brains, and I think it is, it explains why exponential growth takes us by surprise: it sneaks up on us by seeming linear at first. Then it explodes.

Here's a story illustrating this explosiveness.[2] Ernst Stavro Blofeld has handcuffed James Bond to a seat in the top row in the Rungrado May Day Stadium[3] in North Korea, the largest stadium in the world. Blofeld informs Mr. Bond that workers from SPECTRE have waterproofed the stadium and have placed a machine on the center of the field, far below, that releases one drop of water[4] after the first minute, two drops after the second minute, four drops after the third minute, and continues doubling the number of drops each minute. He then says, "Goodbye, Mr. Bond," and leaves. Bond has discerned the make and model of his handcuffs from the sound they'd made when closed; from his extensive practice, he knows it will take him exactly 52 minutes to escape.

Bond feels confident as he works. After all, he can't see any water on the field, and it's just a few drops each minute. *This is too easy*, he thinks. *Blofeld has finally lost it.* After 25 minutes, Bond can barely make out a glimmer far down on the field that might or might not be a puddle. Bond thinks nothing of it and concentrates on the task at hand. When he looks up again 19 minutes later, however, he's shocked to see that the field is covered by water to a depth of 14 meters, and the surface is rising visibly. Bond barely has time to take his final preparatory breaths for extended immersion, because two minutes later he's under water.[5]

This story is an allegory for global warming or any one of a number of interconnected exponential processes to be explored in this chapter. Human greenhouse gas emissions (mainly CO_2) and atmospheric CO_2 concentration are both growing exponentially at a rate of 2.2% per year (see Chapter 3). Something growing at a continuous rate of R percent per year doubles every $69/R$ years.[6] At 2.2% annual growth, the cumulative amount of greenhouse gases we've dumped into the atmosphere is doubling every 30 years.[7]

Before James Watt patented his steam engine in 1781, the atmosphere's CO_2 concentration was 280 ppm. The Earth's climate

TABLE 5.1. CO_2 concentrations at 2.2% annual growth.

Year	ppm CO_2	% of 280 ppm
1793	280.9	0%
1825	281.9	1%
1856	283.8	1%
1888	287.5	3%
1919	295	5%
1951	310	11%
1982	340	21%
2014	400	43%
2046	520	86%
2077	760	170%
2109	1240	340%
2140	2200	690%

was stable and amenable for human civilization. By 2014, the CO_2 concentration had risen to 400 ppm. This increase implies that if we continue business as usual, 30 years after 2014 the human contribution of 120 ppm will have doubled, taking us up to 520 ppm in 2044. If we exceed emissions growth of 2.2% per year, we will of course reach this doubling a little sooner; if we ramp down from 2.2% per year, it will take longer.

Table 5.1 shows CO_2 concentrations at 2.2% annual growth. Notice how slowly the growth starts: it takes *five doublings* just to get above a 10% increase. Most humans weren't concerned by this early growth, myself included. In the year 2000, with my linear brain attending to other matters (making money and finding a mate), I tacitly assumed that humanity had 100 years or so to do something about global warming. I wasn't alarmed, and neither were any of my friends or colleagues.

But as you can see, the early 21st century marks a sea change in growth, with the relentless power of doubling causing human CO_2 emissions to suddenly dominate the biosphere's stable, pre-industrial 280 ppm. *Today, we're poised on the brink of explosiveness.* This is one reason I'm certain that global fossil-fueled industrial civilization will soon end, one way or another.

I doubt that
fossil-fueled industrial civilization
will continue for much longer.

Population

Population growth underpins most facets of our predicament. Like our bodies, the biosphere is a complex system that requires balance between many parts to function well. Humanity is just one part of the biosphere, a part that has become out of balance with the rest of the system.

Here's some data. The top pane of Figure 5.1 shows human population from 10,000 years BCE (at which time the population was between 1 and 10 million) until today.[8] The bottom pane of Figure 5.1 shows human population from the Renaissance to the present, and United Nation projections to 2100.[9] The best exponential fit to data between the years 1500 and 2014 has a growth rate of 1.7%.

The global population growth rate peaked at 2.2% per year in 1963 when there were 3.2 billion people, equivalent to 190,000 new humans per day.[10] By 2014, growth had decreased to 1.1% per year, but because there were now more than 7 billion people, this translated to 217,000 additional people *each day*. (Try to imagine 4.3 births and 1.8 deaths each second.)

The main reasons for the slowing birth rate involve empowerment of women: increased female education levels and increased access to contraception. Correlation between lower birth rates and female education, which can be measured using literacy or enrollment rates, is extremely clear; there's strong evidence for causation as well.[11] When women gain more control over family planning and begin to have career options, families tend to delay childbearing and to have fewer children. In addition, empowering women improves their health and life prospects, as well as those of the children they do have.

Note that the growth rate of CO_2 emissions (2.2% per year) is twice the current population growth rate (1.1% per year). This suggests that global emissions are currently being driven in roughly equal measures by population growth and growth in individual consumption.[12]

Both the US Census and the United Nations expect the world population in 2050 to be 9.7 billion[13] and still growing at 0.5% per

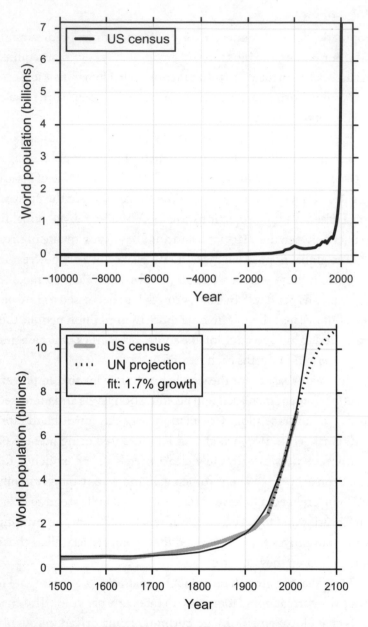

FIGURE 5.1. World population growth. *Top*: from 10,000 years BCE to present. *Bottom*: from the Renaissance to the present, including the UN projection to 2100 and an exponential fit.

year (130,000 additional people per day); in 2100, the UN predicts 11.2 billion[14] with growth of 0.1% per year (30,000 additional people per day).[15] The growth rate is declining, but can Earth sustain 11.2 billion humans?

As we saw earlier, in the short term, we will move away from fossil fuels. For long-term biospherism, we must also settle at a stable and sustainable population.[16] This will be a tall order: biology has given every species an imperative to reproduce, and humans are no exception.

Policies aimed at educating women or encouraging contraceptive use, both in the US and abroad, would help. I'd personally also think long and hard before deciding to bring a new human into tomorrow's world—even longer and harder than I did back in 2006 when Sharon and I decided to have our first baby, and in 2008 when we decided to have our second. We wanted a third, but we decided against it. The global replacement rate is 2.3 children per woman: if no person had more than two children, the global population would steadily come back down to earth.[17]

As we'll discuss below, even the current human population is likely unsustainable. This suggests that, when viewed in the global average, it's actually irresponsible to have more than two children. Of course, I'm not placing the entire onus of population control on individuals and letting institutions off the hook. Nor do I think that a poor woman in rural Bolivia, with no education or access to birth control, is acting irresponsibly by having more than two kids. However, I *am* suggesting that privilege carries responsibility, and that those of us with privilege can choose smaller families with only one or two children, or to adopt. We can also support those who decide not to have children at all. Unfortunately, women who choose to be childless still face social stigma,[18] whereas I believe they deserve respect for their choice.

At the collective level, it may finally be time for the nations of the world to begin addressing global population in some just and decent way, but this won't happen until average people recognize there's a systemic problem. As unpleasant as it is to consider such policy, the only other option—allowing the biosphere to manage our population for us—might be even worse. The biosphere could do this, for example, via continued evolution of viruses targeting developing fetuses. Or it could simply limit our food supply.

Peak food

Food is a prerequisite for population growth, because humans are literally made out of food. In the industrialized world, crop yields per acre began to grow dramatically after about 1950 and in 2015 were at an all-time high. However, that growth came at a high price to our soil, water supply, and atmosphere. Already yield growth is slowing globally and collapsing in some regions.[19] We can't keep growing more food on a finite Earth, year after year, forever. Instead, global food production will reach a maximum value—peak food—and then begin to decline.

> Within a decade, no child will go to bed hungry.
>
> — HENRY KISSINGER, speaking in 1974

The Green Revolution: Eating fossil fuel

By the late 1950s, human agriculture had essentially exhausted Earth's supply of fertile temperate grassland. The global march of grains that began some 10,000 years earlier had run out of new land to plow, so the focus of farming shifted from expansion to intensification: increasing how much grain could be extracted per acre. New industrial farming methods pioneered in the US permeated global agriculture. In 1968, the director of the US Agency for International Development (USAID) named the shift in agriculture the Green Revolution. In 1970, the biologist Norman Borlaug, known as the "father of the Green Revolution" for his work on high-yielding wheat varieties, was awarded the Nobel Peace Prize. But even Borlaug recognized that ramping up food production was only kicking the population can down the road. In his Nobel lecture, he warned, "The Green Revolution has won a temporary success in man's war against hunger and deprivation; it has given man a breathing space.... But the frightening power of human reproduction must also be curbed; otherwise the success of the

Green Revolution will be ephemeral only." Borlaug gave a time scale of three decades.[20]

Four key technologies underpin the Green Revolution: high-yielding crop varieties,[21] irrigation, nitrogen fertilizer, and chemical pesticides. A fifth technology is now pushing yields still further: genetically modified organisms (GMOs).[22]

High-yielding crop varieties are *prima donnas*, prolific only when given pesticides, fertilizers, and irrigation. Resilience has been bred out of them; they focus energy narrowly on seed production. Without pesticides to protect them from insects, without herbicides to protect them from weed competition and to allow dense planting, and without fertilizers to keep their carbohydrate production in overdrive, they're out-produced by the more resilient traditional varieties they were engineered to replace.[23] Between 1961 and 1999, irrigated acreage increased by 97%, and use of pesticides and nitrogen fertilizer increased by 854% and 638% respectively.[24] Today, less than 20% of this nitrogen ends up in our food.[25] The rest becomes nitrate pollution groundwater, washes downstream where it causes algal blooms and dead zones, or is blown into the air where it causes respiratory illness and eventually imbalances forest ecosystems.[26]

Nitrogen fertilizer and chemical pesticides effectively allow growers to treat soil as a dead matrix for holding up plants, turning farms into fossil-fueled food factories. Plants require nitrogen atoms to construct proteins, nucleic acids, and chlorophyll, but nitrogen molecules, which make up 78% of air, have strong triple bonds that render them chemically unavailable to plants. So plants rely on bacteria that are capable of splitting this bond so that nitrogen can be assimilated into amino acids, DNA, and the other building blocks essential to life. Some of these bacteria live freely in the soil, and some partner symbiotically with plants (mainly legumes). This process of making nitrogen available to organisms is called *fixation*.

In 1908, the German chemist Fritz Haber invented a way to use natural gas to split the N_2 triple bond and create fertilizer, opening the industrial age of agriculture. Global fertilizer production, which is fossil-fuel-intensive, grew nearly exponentially until about 1980, at which point the Green Revolution had penetrated most of the world's grain croplands.[27] Today the majority of nitrogen on Earth is fixed industrially; if your diet is typical of industrial civilization, at least one-half of the nitrogen atoms in your body were fixed by the Haber process.[28] In this way, our bodies are indirectly made out of fossil fuel. We are what we eat.

Most other aspects of our industrial food system also run on fossil fuels, from processing to packaging to distribution. For example, in 1997, a pound of produce traveled more than 1,700 miles on average before arriving at a market terminal in Maryland, USA.[29] This agriculture system, while good at producing huge quantities of food (and extracting profits for vertically integrated multinational corporations), is utterly dependent on fossil fuels. Way back in 1991, it took an average of 10–15 calories of fossil energy to produce one calorie of food in the US.[30] One meat calorie required 10–40 times more fossil fuel calories than one fruit or vegetable calorie.[31]

Unsurprisingly, food prices now follow fuel prices in lockstep (Figure 5.2).[32] If fuel prices rise or become more volatile, food prices will follow. Higher prices and higher volatility would both be bad news for the world's poor.

The Green Revolution, growth, and hunger

The Green Revolution was motivated by a vision of ending world hunger. The rapid adoption of intensive fertilization, irrigation, pesticide-supported monoculture, and crop varieties optimized to these industrial methods thereby seized the moral high ground. And indeed it might have ended world hunger—if only humans had shared the food equitably and *stopped making so many new mouths to feed*. But any population of animals responds to the available food supply—and once again, humans are no exception.[33]

Supercharging agriculture with fossil fuels has almost tripled the average world grain yield since 1960. The human population

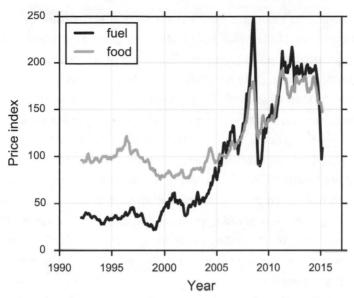

FIGURE 5.2. Global food and fuel price indices. A price index of 100 corresponds to the year 2005.

has increased in lockstep, almost tripling over the same period. And global food demand is expected to increase by another factor of two or three by 2050.[34]

Because humans are made out of food, the size of our population depends on the amount of food available to us. Only a small number of species engage in agriculture and can therefore intentionally increase food production; this group includes ants, termites, damselfish, humans, and others. Our agricultural ability perhaps gives the illusion that we have conscious control over the dynamic equilibrium between our food supply and our population. But so far, industrial societies have *always* chosen to increase the food supply, perhaps both from fear of famine and desire for wealth; and the population has *always* responded predictably, as would the population of any other species—by increasing.

So the Green Revolution increased the human food supply, driving growth of the human population. As a sort of aside, we can ask whether it reduced global hunger. Hunger turns out to be complex and difficult even to define, let alone measure.

According to estimates from the Food and Agriculture Organization of the United Nations (FAO), the number of chronically hungry people in the world has fluctuated near one billion since the 1970s, peaking at over a billion in 1993 and again in 2009, and reaching a low of around 800 million in 2011.[35] Because the global population has increased, the *percentage* of hungry people in the world has decreased, but the actual *number* of hungry people has remained more-or-less constant. We can't exclude the possibility that without the Green Revolution the number of hungry people would have been even greater. Nonetheless, it seems fair to say that we've done a grand experiment, starting in the 1950s, and we've found that throwing more food at the problem of world hunger creates more people, but it doesn't diminish the number of hungry people.

In any case, the fact is that we've been running the global agricultural engine ever faster, creating more humans, more greenhouse gases, and a need for ever more food. This runaway cycle is now facing rising stresses on multiple fronts such as water depletion, soil degradation, and global warming. How long can we keep this up?

Since 1798, when Thomas Malthus published *An Essay on the Principle of Population*, people have been making dire prediction about hitting agriculture's limits, a crisis which has been postponed by technological innovations such as the Green Revolution. Because of the inherent unpredictability of technological innovation, these limits of food production remain unclear. Nonetheless, I hold that there's a physical upper limit on how much food we can produce on Earth, technological innovation notwithstanding; and that we must someday hit a hard biophysical limit to agricultural growth.

In the early 21st century, for the first time, there's evidence that we may actually be approaching this hard limit. Despite our best technological efforts, one-third of global grain production shows evidence of plateaus or abrupt decreases in yield rates.[36] There's discussion now about the need for a "Green Revolution 2.0."[37] But how far can we push this system? And what will happen when we reach that limit?

Furthermore, we now know that these technological innovations come at a high cost. The fossil-fueled industrial food system is not only causing great damage to natural systems and biodiversity, it has also replaced the more resilient and integrated local food systems that predated it—systems that are increasingly perceived as the sustainable and economical path forward for agriculture.[38]

Food system resilience increases with crop diversity,[39] and local food systems utilize a greater diversity of crop varieties and rely more on locally adapted seeds within the community from year to year (as opposed to seeds for a single engineered variety purchased from a multinational corporation).[40] The erasure of local food systems in much of the industrialized world included erasure of grassroots knowledge and infrastructure at the individual level (gardening, canning, keeping chickens, saving seeds, pruning fruit trees) as well as the community level (local farms, cottage businesses, local markets). If the industrial food system does collapse at some point, your survival could possibly depend on whether and to what extent your community has managed to relocalize food production. It will take time to shift to growing food locally and without fossil fuels.

Water depletion

In many regions, humans are pumping groundwater from aquifers far faster than the aquifers can replenish. The Ogallala Aquifer covers a vast expanse of the US high plains from Nebraska to Texas; we're currently depleting it at nine times its recharge rate (a depletion ratio of 9). Other aquifers in the world are faring even worse. The depletion ratio is 27 for the Western Mexico Aquifer, 48 for the Northern Arabian Aquifer, and a whopping 54 for the Upper Ganges Aquifer, which provides water for Northern India and Pakistan, including the subcontinent's wheat belt.[41]

Depletion ratios are well-known, but total water amounts in aquifers are not. If we continue depleting these aquifers, they will someday run dry, but it's difficult to predict when. However, it's possible to make such estimates from measurements of bedrock elevation and predevelopment groundwater level. A recent study of the Ogallala Aquifer estimates that, in 2010, 30% of the water was gone, and that farming in western Kansas will peak around 2040 and then decline due to lack of available irrigation water.[42] This would transform the landscape of North American agriculture.[43]

Extraction isn't the only threat to the world's freshwater supplies; there's also global warming. For example, California depends on Sierra Nevada snowpack both for Central Valley agriculture and for the cities of Los Angeles and San Francisco. The snowpack is predicted to decrease by 70–80% by 2070–2099.[44] In California, as in many other regions of the world, we're living unsustainably on borrowed groundwater.

Global warming is predicted to cause deeper droughts in the future, worse than the drought that contributed to the collapse of the Anasazi civilization.[45] Aquifers are like shock absorbers for drought; their depletion would amplify drought impact. Depletion in key agricultural regions (such as the US high plains, Mexico, India, and northern China) would have far-reaching consequences: no water, no food.

Soil degradation

Soil is amazing, the thin living skin of a planet which would otherwise be a lifeless rock. Industrial agriculture strips the soil of its nutrients, its microbial diversity, its ability to hold water, and its ability to resist erosion. Degraded soils are unable to provide the normal ecosystem services of healthy soils. About one-quarter of the world's agricultural lands are thought to have highly degraded soils,[46] which may cause a significant decrease in crop yields by midcentury.[47]

Good soil isn't compacted, has a healthy tilth (soil structure, which includes pores and aggregates), proper drainage, adequate organic material, and adequate nutrients (both macronutrients like nitrogen, potassium, and phosphorous and micronutrients like manganese). Good soil has few disease pathogens or pest insects, and no toxins. It's a self-organizing system driven by a complex web of plant roots, beneficial bacteria, fungi, and insects,[48] with a delicate, water-retaining three-dimensional structure.[49]

Soil degradation occurs for a variety of reasons, such as soil mining (growing and shipping crops without replenishing organic matter), application of fertilizers and pesticides (which kill soil microbes), tilling (which destroys tilth), compaction, irrigation-induced salinity, monoculture (lack of plant diversity, which affects soil microbes), and erosion. All of these are hallmarks of industrial agribusiness. In addition, urban expansion contributes by paving over living soils.

Degraded soils hold one-half the water of healthy soils.[50] Healthy soils act like a sponge, holding water in the root zone, while degraded soils allow water to rapidly descend below the root zone and out of the reach of plants. Depleted soils require more irrigation.

Soils heal slowly, and it will take a sea change in agricultural mindset in order to allow healing to begin. Humans will need to stop thinking of food as a global commodity, and to start seeing it for what it is: a critical interface between humans and the biosphere, our connection to life. We will need economic systems that support farmers who use biospheric methods to produce high-quality food, because switching to no-till and green manure methods can take a few seasons to pay off financially. The sea change may already be underway: in the US, about 35% of cropland is already farmed without tilling, and the idea is spreading.[51]

Those of us who aren't farmers can help by buying responsibly grown food, by wasting less of it, and by eating less meat. Even better, we can start growing food, thereby developing awareness of the land. We can become growers of good soil. In my experience, it's possible to grow good soil on an individual scale through composting and recycling humanure (see Chapter 12). I suspect that the main reason homegrown food tastes so much better than industrially produced food is because of the soil.

Soil degradation is connected to deforestation and global warming. In many parts of the tropics, agriculture degrades soils, abandons the degraded land, clears virgin forests to farm, and repeats. The net loss of 35,000 acres of forest per day[52] in turn moves carbon into the atmosphere, accelerating global warming.

Global warming

The interaction between climate change and agriculture is complex. Crop ranges are shifting, some northern regions will benefit agriculturally, and farmers will adapt. Overall, though, global warming spells trouble for agriculture.

First, as just discussed, global warming is a threat to the world's supply of freshwater for irrigation.

Second, increasing temperatures and related droughts are already adversely affecting cereal crop yields, and this impact is predicted to deepen.[53] Global yields have already been reduced an estimated 6% to 10% (and yields might continue decreasing 3% to 5% per 0.5°C increase in temperature).[54] Crop failures caused by heat waves and droughts forced Russia to stop exporting grain in 2010.[55] Although corporate researchers are genetically modifying crops for heat resistance, this can only be taken so far: as temperatures increase, eventually sensitive plant molecular systems (e.g., lipid membranes) break down.[56]

Third, pest and disease ranges are expected to expand with warming temperatures. Fungus, oomycete, and insect pests, normally kept in check by cold weather, are migrating northward at the same rate as warmer temperatures.[57]

Fourth, as CO_2 concentrations increase, our food becomes less nutritious, as vitamin and protein contents decrease.[58] This nutrition decrease affects the C_3 photosynthetic plants—95% of the plant species on Earth. (The important agricultural exceptions are corn, millet, sorghum, and sugarcane; most other crops are affected.) And the effect is significant: for example, the protein content of goldenrod pollen has fallen by one-third since the beginning of the Industrial Revolution.[59]

Finally, we can expect greater year-to-year variability in harvests, driven in part by higher variability in precipitation.[60]

These adverse effects will be partially offset by the CO_2 fertilization effect: higher atmospheric CO_2 concentrations lead to higher yields, all else being equal.[61] While the precise magnitude of the CO_2 fertilization effect is still uncertain, we do know it won't be large enough to offset the losses.[62]

When is peak food?

Meanwhile, *demand* for food is expected to increase to 150% of 2010 levels by 2030, and to double by 2050.[63] The demand increase is driven by increases in both population and consumption, and yield projections fall far short.[64] It's not clear that we have the capacity to double global food production. And even if we did, would we be able to continue this pace beyond 2050? Do we have the capacity to quadruple current production?

Given the complexity of the global food system and its interconnections with global warming, resource depletion, economics, politics, and global population trends, it's impossible to predict a date for peak food. However, if the population continues to grow, peak food must occur sooner or later. As we approach this peak, the real price of food will increase as demand outstrips supply regionally, contributing to political instability.[65] When people can't eat, the fabric of a society tears—as was dramatically demonstrated by Venezuela's food crisis, which blew up in 2015.[66]

Many of us in wealthy nations currently take food for granted. But it seems possible that peak food will be an increasingly important factor in the deep rearrangement of globalized civilization.

We're not helpless, though. We can grow food ourselves, in our front yards, in our backyards, in community gardens, and in vacant areas. We can create communities capable of feeding themselves without fossil fuels. We can reduce our meat consumption. In these ways, we can gradually transition our agriculture away from industrial monocultures, and toward organic polycultures on smaller scales. This shift will require more of us to spend some time growing food—a practice I personally find rewarding.

Peak fuel

In the early 2000s, people concerned about humanity's unsustainable business-as-usual path seemed to fall into one of two main camps. There were those more concerned about global warming, and there were those more concerned about global economic collapse due to "peak oil," the end of the era of cheap fossil fuels.

By now, though, it has become clear that global warming is the more urgent and perilous of these two dynamics. As Cristophe McGlade and Paul Ekins wrote in the journal *Nature*, "Although there have previously been fears over the scarcity of fossil fuels, in a climate-constrained world this is no longer a relevant concern."[67] And as we saw in Chapter 4, in order to avoid catastrophic global warming, we'll need to leave at least two-thirds of current fossil fuel reserves in the ground.

> All of my friends, my enemies too
> Live in the shadows of the dirtiest fuel
> Burns the land and its people to ruin
> While wind's always whistling an infinite tune
> — DIRTY PROJECTORS, "Just From Chevron"

Still, today's economy is intimately tied to fossil fuels—all economic production requires energy—and as such it is vulnerable to fossil fuel price swings. Indeed, since the 1940s, oil price shocks were the main drivers of economic recessions.[68] It's conceivable that a fuel price spike contributed significantly to the 2008 economic downturn by pressuring the transportation sector, which slowed growth enough to tip over the subprime mortgage pyramid scheme; and it's also conceivable that in a post-peak epoch, economic growth will reliably trigger fuel price spikes and recession, followed by low fuel prices and a gradual return over several years, in a repeating saw-toothed pattern.[69]

The next two sections describe how our economy might be vulnerable to the fuel supply. The slower we are to transition away from fossil fuels, the more likely these scenarios are to be relevant.

Fracking boom and bust

In the beginning of 2016, oil was below $30 per barrel, whereas in the summer of 2014, it was over $100 per barrel. The main reason for this price drop was overproduction, mainly from the US fracking boom. Fracking was an old technology for obtaining "tight" oil and gas that became economically viable after the price of oil passed $70 per barrel. Fracking continued anyway, even at uneconomic prices, because investors continued giving their money to the frackers.[70]

Over geologic time, oil flows from impermeable rock formations like shale to permeable formations like sandstone. In the US, we've long since tapped out the easy oil in permeable formations; it takes more energy, effort, and money to extract fuels from impermeable ("tight") formations. Hydraulic fracturing (fracking) is a method for obtaining the tight oil and natural gas. However, fracking wells play out very rapidly, typically after only a year or so. This is why frackers drill huge numbers of wells.[71]

National or global production is the aggregation of production from many individual wells, each of which ramps up to peak production and then declines. Figure 5.3 shows graphs of US (top) and global (bottom) annual crude oil production.[72] From 1920 until about 1960, US production boomed as the permeable oil reservoirs were developed and tapped. As these easy reservoirs played out, their replacements were less accessible. The cost to retrieve each barrel of US oil went up until the US could no longer compete on the international market. Production rolled off, peaking in 1970 and then declining. Then, in 2010 fracking kicked in, creating a dramatic uptick in production.

In the 1950s, M. King Hubbert developed a mathematical theory for the extraction of finite, non-replenishing resources: production ramps up initially, reaches a peak, and then ramps down as the resource becomes increasingly difficult to obtain. Non-intuitively, the ramp-down happens at the same rate as the ramp-up; the faster production rises, the faster it declines post-peak.[73]

Given our globally connected market for energy, the global economy will of course respond to the global peak. Nonetheless,

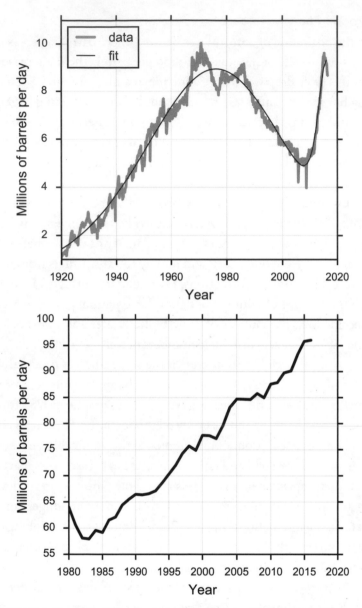

FIGURE 5.3. *Top*: US crude oil production. *Bottom*: Global crude oil production.

I found it informative to fit the US data with the sum of two independent Hubbert curves.[74] One Hubbert curve fits the conventional crude oil production, which peaks around 1970, and the other fits the more recent fracking production. The Hubbert curves fit the data remarkably well.

No one knows exactly how long the fracking boom will continue, but one report predicts US peaks in both tight oil and gas production before 2020.[75] The US government agrees that tight oil production will peak by 2020 "as drilling moves into less productive areas."[76] On the far side of the production peak, Hubbert theory predicts rapid collapse which may cause oil and gas price increases, destabilizing the economy.

As for global peak oil, the consensus among experts is that it will occur sometime between now and 2030. Indeed, it may have occurred already.[77]

Fossil fuels are increasingly costly to extract

It takes energy to obtain energy: you need fuel to make equipment, drill wells, fracture layers of rock, and refine crude products. The energy returned on energy invested (EROEI) is the ratio of how many units of energy you get per unit of energy spent.

Fossil fuels are becoming more difficult to obtain, and EROEI for fossil fuels is declining rapidly. The EROEI for coal in the US declined from 80:1 in the 1950s to 30:1 in the 1970s.[78] Globally, the mean EROEI of oil and gas halved in just ten years, from 40:1 in 1995 to 20:1 in 2006.[79] Extracting oil from tar sands is even more energy-intensive: tar sands oil has an EROEI of only about 4:1.[80] This is part of a longer global trend of decreasing oil and gas EROEI, which is likely already having a large impact on the global economy.[81]

Declining EROEI increases the real cost of energy and creates a drag on the economy. Money is essentially a social contract for storing work, and fossil fuels equal work. This essential equivalence between fossil fuels and money is partly why we're finding it difficult to leave them in the ground.

TABLE 5.2. EROEI estimates for various energy sources.

Source	Year	EROEI
Oil and gas (global)	1995	40:1
Oil and gas (global)	2006	20:1
Coal (US)	1950s	80:1
Coal (China)	2010	30:1
Tar sands		4:1
Hydropower		100:1
Wind (w/o storage)		20:1
Nuclear (US)		10:1
Solar PV (w/o storage)		10:1
Corn ethanol		1:1

Table 5.2 gives published EROEI estimates for several energy sources.[82] EROEIs of renewable generation technologies like solar and wind are low relative to fossil fuels. The numbers here do not include energy storage systems, which would further lower the overall EROEIs for renewables. This doesn't necessarily imply that our renewably powered lives will be bleak in the future: EROEIs even as low as 3:1 can still provide useful energy to society. But it does suggest that our current systems of energy consumption and production, and therefore our economic systems, will undergo a tectonic shift. The age of nearly free energy appears to be drawing to a close.

Indeed, both the world and US GDP growth rates have fallen significantly over the last 60 years.[83] No one knows the exact reasons for this slowdown, but three possible causes are deceleration in population growth, deceleration in labor productivity growth or technological growth, and the decreasing EROEI of fossil fuels. I suspect that the slowdown indicates major cracks in the foundation of the industrial economy: its energy and money systems are becoming increasingly unsustainable.

The money vortex

Apparently capitalism has a structural flaw: money exhibits a gravitational attraction whereby wealth accrues more wealth. The debt-based money system ensures this is so, via interest payments on use of capital. The inevitable result is a black hole of wealth that warps the structures of power, accelerating the process.

When an individual decides to engage in a capital-intensive activity (such as extracting oil from the Earth and refining it into fuel), he or she will find that forming a corporation brings significant advantages. First, the individual is no longer liable for debts or misdeeds; this liability transfers to the corporation. Second, a corporation leverages the wealth of a potentially enormous pool of investors. Third, unlike a human, a corporation can exist perpetually.

Once formed, corporations seek to make as much profit as possible— indeed, they're legally bound to do so.[84] Naturally they seek to pay as little as possible in taxes and to limit any regulation that might impinge on profits. Their only goal is to grow as rapidly as possible.

> My philosophy is to make money.
>
> — REX TILLERSON, former CEO of Exxon Mobil Corporation

The politicians who establish taxes and regulations want above all else to be reelected. But effective campaigns are expensive. So the corporations and their wealthy owners offer money; in exchange they ask for changes to laws, regulations, and tax codes. The politician who doesn't go along will lose that source of funding, decreasing her odds for reelection.

Corporations, then, continuously reshape the legal landscape so as to exert more influence and thus extract more wealth. There are innumerable real-world examples, including *Citizens United*, a 2010 US Supreme Court decision allowing corporations unlimited funding access to political campaigns;[85] the American Legislative Exchange Council (ALEC), a group of corporations and lobbyists that literally writes laws and then gets state legislatures to pass them;[86] and international trade deals which exert downward pressure on wages and allow corporations to dismantle national regulations that interfere with profits. As voters, much of this corporate influence is out of our control: the two-party system in the US, for example, has long ensured that almost every candidate with a reasonable chance of winning high office is under corporate influence—Republican and Democrat alike.

This infiltration of the political system brings capital ever more wealth, growth, and control over the halls of power, a positive feedback loop. Systems with such a feedback are inherently unstable. (The agriculture-population system is another example.) With very vigilant control, we can prevent this feedback with checks and balances; but as soon as there's a lapse in our vigilance, the feedback—wealth's self-gravity—will again metastasize. We can call this metastasis the *corporatocracy*. Its building blocks are corporations, wealthy individuals, politicians, and laws. Its essence is the systematized love of money.

Climate stalemate

Fossil fuel corporations are among the largest and most profitable corporations the planet has ever seen,[87] and all industrial extraction of wealth, from agriculture to mining to manufacturing, runs on fossil fuels. Not surprisingly, the corporatocracy views climate action as a threat to its survival. To preserve the *status quo*, it actively blocks meaningful climate action by controlling policymakers, and by confusing the public by falsely sowing scientific doubt.[88]

Effective climate action is certainly hamstrung under corporatocracy, which prioritizes rent-seeking over the biosphere. Even without the additional burden of runaway corporatocracy, modern capitalistic democracies would still tend to prioritize economic growth over the biosphere. For example, a 1992 United Nations international climate agreement explicitly says that "measures taken to combat climate change, including unilateral ones, should not constitute…a disguised restriction on international trade,"[89] and these priorities have held ever since. Global extractivist trade, and the growth and consumption that are its lifeblood, appears incompatible with meaningful climate mitigation. This is why we see efforts, for example, to put price tags on ecosystem services. I personally fear that these sorts of incremental efforts are too little, too late; and that what we really require is a paradigm shift: instead of viewing the biosphere as part of the human economy, we need to view the human economy as part of the biosphere.

Living without growth

Our industrial society is addicted to exponential economic growth. When growth is down, suffering ensues. People lose jobs, wages stagnate, and the specter of a systematic Great Depression-like unraveling hangs in the air. The need for growth is locked into our political-economic system: politicians need to promise exponential growth to get elected. Some economists even argue that exponential growth is locked into the DNA of our economy at the level of the debt-based money system, which requires growth to service interest payments on debt.[90]

And this economic growth requires ever more consumption of energy, habitat, and other natural resources. This has been the case historically, and it's possible to show that economic growth can never permanently decouple from natural resource use.[91] We are embodied, physical beings, and we'll continue to rely on an embodied, physical economy to meet our needs.

> Anyone who believes exponential growth can go on forever in a finite world is either a madman or an economist.
>
> — KENNETH BOULDING, economist and pioneer of systems theory

However, no law of physics requires exponential economic growth. We humans are free to organize our societies and meet our needs in other ways. Indeed, most human societies over the course of human existence were not organized around exponential growth.

I believe that humanity's grand challenge goes beyond global warming. Responding to global warming is an urgent first step. But in order to avoid repeated cycles of unnecessary suffering, of growth and collapse, humanity must somehow move into a stable and harmonious long-term relationship with the biosphere.

First and foremost, we'll need to learn how to see ourselves as just one part of a vast, complex, and beautiful tapestry of life. We'll need to respect and value nonhumans just as much as humans. We'll need to come out of the hubristic mindset of having dominion over nature. As I said earlier, we'll need to realize that our economic system is a part of the biosphere, and not the other way around. If something we do collectively harms the biosphere, we'll need to stop, no matter what some corporations or individuals might prefer.

This in turn implies that we'll need a just and equitable society, free from the tyranny of a wealthy few. Our current system's existence depends in part on a story to mollify the poor and prevent an uprising: that after some more growth, everyone will be rich; that economic equality for all is just around the corner, if only we run our engines of extraction ever faster. In a non-growing (steady-state) society, this story obviously makes no sense.

We'll need to learn to measure our quality of life in terms of well-being, not money. We'll need to reduce our population and learn to keep it steady. And of course, we'll need to limit our consumption of resources to within the biosphere's capacity for regeneration.

These are huge changes. They require us to overhaul the deepest levels of our social structures and our collective story, the ways in which we construct meaning. This sea change could only occur if enough of us decide it's the way to go, if we're willing to let go of our modern aspirations for convenience and profit above all else, and if we implement it gradually but deliberately. Table 5.3 suggests some of the necessary changes.

I believe that a steady-state economy holds the potential for a better and more satisfying life for all, under a social system that could be sustained indefinitely. But it's very difficult to see how to get there in the real world via real policies. Policymakers might partially choreograph this huge socioeconomic shift intentionally, if compelled to do so by a grassroots groundswell; but it might also come into being when the current system is reorganized, perhaps unintentionally and catastrophically, in response to forces arising from the limits of the biosphere.

TABLE 5.3. Changes needed for a steady-state economy.

Exponential growth economy	Steady-state economy
Biosphere seen as part of economy	Economy seen as part of biosphere
Growth and collapse	Sustainable and regenerative
Money = quality of life	Well-being = quality of life
Money system based on debt	Local reciprocity
Takes biosphere for granted	Values biosphere services
Biosphere seen as tech-replaceable	Biosphere seen as irreplaceable
Externalizes biosphere destruction	Policies protect biosphere
Global extraction	Local production
Short-term view	Long-term view
"Man has dominion over nature"	"We are one among many species"
Jobs replaced by machines	Labor is valued
Makes a few people obscenely rich	Social justice is woven in

Are there too many humans?

The Earth's biosphere is marvelously abundant, capable of supporting billions of humans. It has given birth to a dizzying variety of species in wonderful and unimaginably strange forms, a psychedelic menagerie. But the biosphere's abundance depends on balance and diversity.

The biosphere supports us by providing food, water, oxygen, waste remediation, a stable and temperate climate, evolved protection from diseases, and protection from harmful solar radiation. These life-support services are interconnected. For example, you can't grow food without water; and the distribution of water in space and time depends on the climate. When the human population reaches a size at which any one of these life-support provisions falls below a sustainable level, it has exceeded carrying capacity.

Given this interconnection, and all we've discussed in this chapter, I think the question of the Earth's carrying capacity boils down to this: can we sustain the amount of food we grow today indefinitely into the future?

> The twentieth century has been marked by a profound historical development: an unwitting evolution of the power to seriously impair human life-support systems.
>
> — GRETCHEN DAILY and PAUL EHRLICH, 1992

The Earth's biosphere has given birth
to a dizzying variety of species in
wonderful and unimaginably strange forms,
a psychedelic menagerie.

By this measure, it may be that we have already exceeded the Earth's carrying capacity. It seems likely to me that global food production will be increasingly challenged by aquifer depletion, drought, heat waves, extreme weather, and soil degradation. Our technology allows us to capitalize Earth's ecological services more easily, but it's not a *substitute* for those services. Rising fuel and fertilizer costs will be an additional constraint when the fracking boom ends.

Loss of biodiversity provides a second, independent line of evidence that human impact has already exceeded Earth's carrying capacity. Diversity is a good measure of an ecosystem's health. Today we're in the midst of the Earth's sixth mass extinction. There were half as many wild animals (vertebrates) alive in 2014 as there were 40 years earlier,[92] and today's extinction rate is about 1,000 times higher than the background rate.[93] Humans and livestock account for over 97% of land vertebrate biomass, whereas wild animals account for less than 3%.[94] We have essentially replaced the Earth's wild places with agriculture, the Earth's nonhumans with humans.

In addressing extinction, many conservation efforts focus on individual species. But we also need to think from a systems perspective. Warming and habitat loss exert pressure on entire ecosystems, which can manifest as pressure on individual species through interconnected processes. For example, moose calves in Maine and New Hampshire are killed by winter ticks, which have exploded due to warming.[95] For another, over half of primate species are facing extinction due to the combined pressure from expanding palm oil and rubber plantations, and bush meat hunting.[96] Nonhuman animals are quite literally under relentless, systematic attack from a mechanized and militarized global economy of nearly eight billion humans. Focused conservation efforts might slow the hemorrhaging; but ultimately, the only way to avoid the sixth mass extinction will be to address the underlying causes.

If eight billion humans (living and eating as we're actually living and eating today) is beyond carrying capacity, what's the sustainable limit? No one knows for sure, but we can make an educated guess. For a lower limit, it's likely greater than one billion. At the start of the Industrial Revolution in the 18th century, when all the coal, oil, and gas deposits were still underground, the human population was just under a billion. Agriculture functioned on sunlight, water, and manure—not on the Haber process. Although there were *regional* population collapses before the 18th century,[97] globally the biosphere was carrying one billion humans while remaining in balance. Humans hadn't changed the atmospheric composition yet, fish were still plentiful in the oceans, the bison were in great abundance on the North American plains, and (perhaps most importantly) the extinction rate hadn't yet started its exponential rise. So Earth's carrying capacity is likely between one and eight billion. I'd guess the midpoint, around 4.5 billion, give or take a billion.

What do the experts say? According to one estimate, the Earth's carrying capacity is about 4 billion.[98] This estimate was made by considering the area of land (at average global primary productivity) required to support one globally average person, including ecosystem services. According to another estimate, the 1994 global population of 5.5 billion "clearly exceeded the capacity of Earth to sustain it," and a population of 1.5–2 billion people would be "optimal."[99]

If everyone on the planet became vegetarian, we could likely double the number of humans that could sustainably ride on spaceship Earth.[100] Because about one-third of all food is wasted globally,[101] we could in principle increase carrying capacity by another 25%—about one or two billion people, depending on whether we're meat eaters or vegetarians—by eliminating half of this waste.

So far, this discussion presupposes the existence of agriculture. Author Jared Diamond has argued that agriculture opened a Pandora's box of human misery, and that we might have been better off without it.[102] As we saw above, the pre-agricultural population ten thousand years ago was no more than ten million humans. By applying technological advances to hunting, gathering, and forest gardening, the non-agricultural carrying capacity would probably be in the hundreds of millions. To reach a sustainable carrying capacity in the billions probably requires agriculture.

But now we can see that if we choose agriculture, we have a deep responsibility to somehow exercise restraint in how we share the biosphere. As the Green Revolution has demonstrated, unstable population growth has so far been intrinsic to the intensive agriculture that we practice, which is brutal to nonhumans in its path and therefore to biodiversity; author Daniel Quinn calls it "totalitarian agriculture."[103] The problem here is not just that our population will collapse when we've exceeded our ecological bounds, it's the risk that our totalitarian agriculture will drive the sixth mass extinction in the process, irreparably impoverishing the biosphere for the next ten million years. Therefore, a sustainable global agriculture requires either a population cap or a land-use cap.

Translating this knowledge into action

A popular environmental slogan is "Think globally, act locally." However, I find it difficult to truly grasp global scales. It can feel overwhelming, even paralyzing. I prefer instead to think locally as well as to act locally, in terms of *human* scales: a garden, a bicycle ride, a community event.

But our predicament is, in fact, global; and it's important to understand it globally in order to forge an appropriate personal response. Global warming is increasing exponentially: I can begin to systematically reduce my own fossil fuel use and support local, state, and national measures that reduce fossil fuel use. A global economy based on exponential growth is eating the planet alive: I can reduce my own consumption and help build a satisfying stable local economy. Global population is driving all of the above: I can choose to have a small family.

Naysayers claim that our predicament is so big that individual actions don't matter, thereby justifying their inaction. I disagree, of course; this viewpoint may even be quintessentially evil, if the only thing necessary for the triumph of evil is indeed for good people to do nothing. The naysayers and I disagree fundamentally on the nature of connections between individuals, and feedbacks between individuals and society.

I believe that social change emerges complexly from resonant movements between many individuals. Today, an increasing number of people are looking beyond green consumerism, beyond political parties, to search for ways to live in alignment with the biosphere. This movement, *biospherism*, resonates with me. I hope this book gives it momentum.

We need to respond, and quickly. We've recently entered an epoch of viscerally evident physical changes due to global warming. As the perception of our predicament goes mainstream, I suspect it will trigger social change. But because fear will be one driver of this change, not all of what evolves will be positive for human beings. Grassroots movements such as biospherism can perhaps create channels for directing the coming social change in a positive direction.

Every day is precious, especially in light of our exponentially unfolding predicament. I hope that you'll join me in using these wonderful days to begin aligning with the Earth, with each other, and with yourself. As more of us change to live in this way, the resilient communities we'll need to weather the coming storms will emerge.

CHAPTER 6

Our Mindset

We have met the enemy,
and he is us.

— WALT KELLY

The root of our predicament lies in our own minds.

Together, our myths, paradigms, and mental habits make up our *mindset* or worldview, a mental edifice that exists at both the individual and collective levels and functions as a link between them. Our mindset feeds back complexly with the entirety of the physical human world, shaping (and being shaped by) our systems of transportation, economy, food, education, community, entertainment, and warfare.

Our mindset tends to bias our assessment of our predicament, leading us to underestimate its depth and urgency. In addition to limiting our perception of the present, it limits our capacity to imagine possible futures. An effective response to our predicament must begin with our mindset. While it is not easy to perceive and change one's mindset, it is possible, with dedicated practice.

The myth of progress

When I was a kid in elementary school, I spent hours playing *Star Trek*. My best friend pretended to be Kirk, and I pretended to be Spock (and my sister's cat was a respected admiral, a "Catenesian"). For hours at a time, we zoomed around the galaxy at warp factor nine, exploring exotic planets inhabited by Romulans, green alien women, and other dangerous things. Our amazing

23rd-century technology saved us every time. We were playing the *myth of progress.*

A myth (Greek *mythos*, story or word) is a primal story that a society uses to understand its origin and relationship to the rest of the world. Our experience of myth is transparent: we aren't typically aware of its influence in daily life, although that influence is profound. Myths help us make sense of the confusing world into which we're born, and a challenge to them creates existential anxiety. We hold myths to be unquestionable and self-evident, and if they do happen to be challenged, we react defensively and emotionally. Societies or groups can have difficulty accommodating ideas that challenge their key myths.[1]

Myths can also make it more difficult to see alternatives to the way things are. As the French literary theorist Roland Barthes wrote, "Myth has the task of giving an historical intention a natural justification, and making contingency appear eternal."[2] We tend to believe that the way things are is how they must be. This is of course false—things could *always* be different—but such belief is a tremendously strong force, and of critical relevance to our predicament. Donella Meadows, lead author of *Limits to Growth*, wrote:

> The shared idea in the minds of society, the great unstated assumptions—unstated because unnecessary to state; everyone knows them—constitute that society's deepest set of beliefs about how the world works. There is a difference between nouns and verbs. People who are paid less are worth less. Growth is good. Nature is a stock of resources to be converted to human purposes. Evolution stopped with the emergence of *Homo sapiens*. One can "own" land. Those are just a few of the paradigmatic assumptions of our culture, all of which utterly dumbfound people of other cultures.[3]

Myths are just as prevalent in modern industrial society as in so-called primitive societies. Indeed, the myth of progress informs us[4] that our industrial society is special precisely because we've finally transcended such primitive thought systems as myth and superstition.

But the myth of progress—a deep, subconscious belief that we're more advanced than past peoples, and that in the future we're certain to be even more advanced—has, ironically, become our most important modern superstition. Of course, this myth sounds thoroughly modern to our ears. For example, we might believe that They (the supernatural entity invoked by the incantation, "They will think of something") will solve global warming with artificial photosynthesis, or solar panels, or thorium reactors. Whether or not this ends up being the case, our *certainty* that They will do this strikes me as superstitious, a kind of magical thinking or blind faith.

Indeed, progress has become a civil religion—a non-theistic belief system that provides the psychological benefits of religion, in this case by making technology sacred. We speak of technology saving us, and we have blind faith that it will. Some of us imagine a merging of man and machine in the not-so-distant future, in which we'll live forever and fly without limits through the galaxy; some, that our salvation lies in leaving a dead Earth to colonize other planets. Is this a technological conceptualization of heaven? There's a striking similarity between the iconic image of a fiery rocket rising into the heavens and the ascension of Christ.[5]

We find the assumption of perpetual social, economic, and technological progress reassuring, a sign that we're on track as a species, that our contributions at work mean something. Either consciously or unconsciously, we extrapolate this progress to a future techno-utopia, allowing us to view today's problems as temporary and justifying our too-often unsatisfying roles in society. Experts who make the case for perpetual progress tend to be popular, because their message resonates with our comforting myth. Of course, there's no guarantee of perpetual progress. In the cycling of civilizations, there have been dark ages in the past. Who can say there won't be dark ages in the future?

Over the last few hundred years in particular, though, technology has indeed progressed steadily. We've divided knowledge into focused disciplines, the singular domains of specialists. No one can begin to understand how it all works, and we don't need to. After all, when my son Zane pricked his foot on a rose as a toddler and developed a life-threatening infection, I didn't need to know *how* antibiotics work; I just needed to know *that* they work. Ironically, this state of affairs means that we tend to view the technological specialists—scientists, doctors, digital gurus—through the lens of our superstition. Those experts are our shamans, in white lab coats.

And when actual science and the myth of progress collide, the myth wins. When an offering of science aligns with the myth—when science feels like progress—our culture celebrates it.[6] But when an offering of science conflicts with the myth, our culture is likely to reject it.

Take global warming, which implies that our technological and economic growth are damaging the biosphere. In this case, science is telling us that technological and economic growth, far from lighting the way for future generations, is actually destroying their prospects. That somehow, fundamentally, human progress is threatening the totality of life on this planet. This message is, of course, deeply *anti-progress*; which helps to explain why something as scientifically clear as global warming[7] can become politicized and challenged, especially by non-scientists.

Technology's proper place

The word *technology* comes from two Greek words and means "the systematic treatment of arts and crafts." Technology can be defined as the development and use of techniques and tools by an animal, with an intention to facilitate the relationship between the animal and matter (the animal's environment). Technology can be used to procure food, security, and comfort. It can be used to make music or to record and remember information.

Many human technological innovations have been, and will continue to be, incredibly useful. I'd even go so far as to suggest that human survival depends on technology, and has done so since the advent of our species. Through the many twists and turns of evolution, our species ended up not with fur and powerful claws, but with community and technology. This has been a remarkably successful strategy—at least until now.

Today, however, the *scale* of human technology has clearly begun to influence the physics of the Earth system. The hundreds of millions of internal combustion engines running at this very moment are smothering the biosphere in their exhaust gases. In this sense, we need *less* technology, not more.

Furthermore, we can be selective in the technologies we choose to invite into our world. Much technology is worth keeping, but perhaps not all. We especially fetishize technologies that are talismanic of the myth of progress: for example, 3D printers, the Internet of Things, social media, virtual reality. Do these technologies truly make us happier? What about self-driving cars and voice assistants? Is this the world we really want to live in, or are there perhaps more interesting and kinder dimensions to explore? I'm not necessarily saying no to these technologies, only that we might want to question them more.

We also need to realize that our survival depends not just on technology, but on community as well. And many of the technologies of petroculture replace or displace direct human interaction, strengthening the illusion that we are separate from each other, the illusion that it's possible for anything to exist separately from everything else.

The myth of separation

A formative moment for me as a teenager growing up in suburban Chicago occurred while I was out walking one night. No one else was outside. Occasionally a car would swoosh by. From every house came the eerie blue synchronous flickering of television.

Something important was lacking, although at the time I wouldn't have been able to articulate what it was. Suburban alienation unconsciously permeated my teenage years.

We construct lives for ourselves that maximize our separation: houses with fences in suburbs; cars; office cubicles; neighbors who don't know one another's names. The money system isolates us from each other. People disappear into television and video games for hours at a time. The American Dream itself can be viewed as the quest for perfecting this separation: once we make our millions, says the Dream, we can live at the top of our skyscraper and fly everywhere in our private jet.

All of these stories emerge from a second fundamental myth of industrial society: that each of us is a separate being in competition with everyone else. This zero-sum mindset colors our interactions with each other, with nature, and even with ourselves. It leads to anxiety, loneliness, and divorce. It leads to placing profit and convenience above all other values. It leads to fear, hatred, and violence.

Of course, at the superficial level, we seem to be separate from everything else. But upon closer examination, this apparent separation dissolves. The deeper reality is that we are connected by society, by our relationships, by the food we eat, by the air we breathe, and by the biosphere that we share—in myriad irrefutable, physical, bodily ways.

Tribalism

Groups of humans with a common identity—whether tribes, political parties, nations, races, or religions—tend to feel separated and afraid of other groups. We humans use mechanisms such as language, clothes, stories, and beliefs to identify individuals in our group and to set ourselves apart from others. Our need to belong to a group is so strong that we tend to take on all the beliefs of our group. Open-minded, fact-based thinking shuts down.

We use stereotypes to make those in other groups seem less than human. If our group has more power, we strip the others of rights, access to resources, and privilege. We encode their lesser status within our legal system, which lowers the threshold for violence against them, and institutionalizes the violence.

We humans create narratives in which people we see as unlike us are evil, so that if we can eradicate them, we'll have eradicated evil. And so the myth of separation has the power to allow us to carry out atrocities; we may even think we're doing good. This is how easy it is to lose our way when we're afraid. The myth of separation distorts community into tribalism.

These tribal narratives have caused murder and suffering throughout history, and still do. Both nationalism and religion can readily create the mindset for war, when acts of violence can come to seem virtuous to their perpetrators. When the mind is closed, true kindness and compassion are impossible.

While I prefer a diversity of cultures, languages, and ways of life on this planet, I'm also dismayed by humanity's continuing inability to coexist peacefully. We are difficult animals; we walk around with these ridiculous, fearful egos that react and escalate conflict. This happens all the time, both in daily life and on the world stage. I'm certain we could do better, if we made a *choice* to do better.

Conquest of nature

Within industrial society, we have a tendency to view nature as something to be conquered. We fear that if we fail to conquer nature, it will kill us; we dream that if we succeed, we can make vast fortunes.

The mentality of conquest has helped us make great advances in medicine. But as a paradigm, it may no longer be humanity's best bet for good health. For example, health experts warn that we've entered "the end of antibiotics" as multi-resistant bacteria out-innovate us.[8] Resistant fungal infections are also on the rise due to widespread use of antifungals in agriculture.[9]

And God said unto them, be fruitful, and multiply, and replenish the earth, and subdue it: and have dominion over the fish of the sea, and over the fowl of the air, and over every living thing that moveth upon the earth.

— GENESIS 1:28

This is also the case in agriculture. Over the course of the 20th century, industrial agriculture increased yields tremendously. But now we know that totalitarian agriculture obliterates complex ecosystems, creating damage that ripples through the biosphere. Now we know that industrial livestock production creates the potential for zoonotic pandemics. The totalitarian agricultural mindset may not be in our best interest, after all.

We approach these confrontations with nature much as we approach warfare, by bringing in more firepower. But ultimately, we're at war with ourselves. Instead of fruitlessly trying to conquer nature, we could be dancing with the biosphere.

Human exceptionalism and nonhuman intelligence

There are likely upward of a trillion distinct species on Earth.[10] Each species arises, thrives for some time in its ecological context, and eventually passes away. While here, each species exhibits its own unique intelligence for survival.

Among this multitude of species, only one is foolish enough to do damage, on a global scale, to the biosphere that gives it life. What's more, this species knows it's doing this damage, and is therefore *choosing* to do it; and this species, strangely, thinks of itself as by far the most intelligent species on the planet, and uses this as a moral excuse to treat other species brutally. This bizarre species, of course, is *homo sapiens*, "wise man."

Not every human culture has believed that humans are exceptional among species; many indigenous people did not, and do not, share this belief. For example, aboriginal people in Australia believe that nonhuman animals and plants were once people, an expression of oneness between beings.[11] I believe that the modern human exceptionalism central to the industrial mindset is a driver of our current ecosocial predicament.

An appropriate response includes a paradigm shift toward recognizing that humans don't have a monopoly on intelligence, and that our inability to recognize and respect nonhuman intelligence is self-limiting. Perhaps our technology has outpaced our wisdom. Perhaps we need gorillas, whose intelligence is so similar to ours, who care for their infants and children just as we do, who we find so easy to identify with—to help us remember respect for the peculiar intelligence of every species. Bees have their own kind of intelligence; so do bacteria in the soil.

The biosphere has a wisdom of its own. The Earth knows the way.

Habits of our lizard brains

In addition to myths, several built-in tendencies of our brain also make it more difficult to see reality as it is. Mitigating global warming is a wicked problem, in part because global warming seems almost perfectly designed to defeat the human brain.

Immediacy

Our brains naturally respond strongly to dangerous situations, like a growling dog or the rent deadline, that are right in front of us. We respond less strongly to future or distant dangers, because we don't *feel* them as threats. We react to the visceral, adrenal sensation of immediate fear more than to intellectual understanding of danger. This discounting of the future explains why we tend to procrastinate: the future feels safely far away, and we don't understand how one moment flows into the next and the future suddenly becomes the present. Then, right before the deadline, we ask ourselves, "where did all the time go?"

Global warming lacks immediacy for most people. Even becoming well-informed doesn't necessarily produce immediacy. For example, I know several astrophysicists who are well-informed about global warming, and who also have large climate footprints due to their frequent flying. When I independently asked three of them how they rectify this apparent contradiction, I was surprised when they all gave me the same reply, nearly word-for-word: "I just don't think about it."

Interestingly, there's a striking difference in the social dynamics of fossil fuel use and water use in Southern California. Wasting water isn't socially acceptable. During a drought, people get very angry about it. On the other hand, burning fossil fuels is not only socially acceptable, it's socially rewarded. Frequent fliers post trip photos on Facebook, enjoy career advantages, and are seen as successful. This difference might be partly because water is more immediate than global warming.

Confirmation bias

We have a natural tendency to look for evidence that supports our beliefs and to ignore evidence that contradicts those beliefs. This is known as *confirmation bias*, and it helps our brains make sense of an enormous volume and complexity of information. Once we've developed our worldview, confirmation bias informs what we choose to believe or disbelieve. If we have a worldview in which global warming doesn't make sense, confirmation bias leads us to prefer evidence that supports the hypothesis that global warming doesn't exist or isn't a problem. In this way, without realizing it, we manage our worldview.

When I talk to people who deny global warming, I usually get a sense that they *need* to believe that the planet isn't warming. Perhaps global warming challenges their worldview at a religious level, or a mythic level: humanity's purpose is to conquer nature in the name of progress. Perhaps denial allows them to avoid feeling guilty about their lifestyle. Or perhaps they simply find global warming so frightening that they have a psychological need to deny it. In any case, their reasons run deeper than intellect.

Optimism bias

Our brains tend to err on the side of being overly optimistic.[12] This is especially true when we're considering something with low immediacy, or that we view as uncertain. Optimism bias implies that global warming is worse than the average person thinks.

Conformity and normalization

The seminal Asch conformity experiments in psychology demonstrated how individual beliefs and actions depend on a majority group.[13] In the classic experiment, eight college students were shown a drawing of a line segment, and asked which of three other line segments (labeled 1, 2, and 3) was the same length. The three candidate segments were of different lengths, with one very obvious match. The students took turns giving their answers. Seven students were actors, but the eighth student didn't know this; the actors colluded with the psychologist to give the same wrong answer. The eighth student—the actual subject—was then asked to choose. Seventy-five percent of subjects were swayed to the actors' wrong answer at least some of the time, while only 25% of subjects were never swayed.[14] If one of the seven colluding students acted as a "true partner" and gave the correct answer, the subject was more likely to do so as well.

When I learned of this experiment, I remembered how, in 2006 when I first woke up to the truth of global warming, I felt afraid to speak out. At the time, I heard almost no one else speaking out, but I saw everyone else living their lives as if nothing was wrong. I saw no one attempting to use less fossil fuel, so neither did I. What a tremendous force for conformity! I also wasted several years searching for a "true partner" before eventually realizing that I had to do what I could on my own. (I now have many "true partners.")

Although we may know that it's happening, global warming isn't yet terribly immediate for many of us. Meanwhile, everyone around us continues burning fossil fuels like there's no tomorrow, including most of the biggest climate activists—even more normalization. The brute fact of the normalcy of fossil fuels is constantly in our faces. So at some deeper level, we think everything must be OK, despite our intellectual understanding to the contrary. And then the myth of progress offers a convenient psychological bridge, a means to resolve this psychological conflict, in the form of "They'll think of something." The result is no action: we tend to go along with the fossil-fuel-burning crowd.

It's difficult to think of burning fossil fuels as harmful or wrong when everyone around us burns with abandon, every day. It's difficult to even imagine living in some other way. Perhaps this explains the stunning failure of our leading global warming activists to "walk the talk." Perhaps this is a major reason why global warming poses such an unprecedented challenge: our lives in industrial society are so entwined with fossil fuel that it's difficult even to envision what a new mode of living would look like, let alone to live it. Of course, we lived without fossil fuel a mere two centuries ago, but the myth of progress blinds us to this truth.

This is why it's critical we begin saying that burning fossil fuels is causing real harm and needs to stop. It's even more important to begin living this message. Through non-conforming behavior, we cause others to question the prevailing perception of what's normal. In the language of the Asch experiments, we become true partners, those who speak the truth and make it easier for others to do so as well.

Taking the biosphere for granted

What causes some people to want to "drill, baby, drill?" Desire for profit, certainly. But perhaps not all of these people, maybe not a single one of them, would actually trade the biosphere for ten trillion dollars. For of course, without a biosphere, in the cold vacuum of space, even our trillionaire would die within a few seconds.

At some level, these people realize they need the biosphere. But at the same time, they're taking it for granted. The inestimable value of a breath of air goes unnoticed. To my mind, this inconsistency reveals a clear need to rethink the entire concept of wealth (and, conversely, of poverty). To live in a healthy biosphere is absolutely a kind of wealth—perhaps the most real kind of wealth there is, from an astrophysical perspective.

I see Earth's biosphere as a fragile and miraculous entity. It's amazing to me that such a thing came to be: this extraordinarily rich but razor-thin layer of interdependent life surrounding a rock floating in the cold vastness of space. How strange! There's no guarantee that Earth's biosphere will remain so richly support-ive of human life. The Earth system answers only to the laws of physics, not to the needs of humans.

Plain old ignorance

Few people have a clear sense of the relative ecological costs of their actions. Many are the avid environmentalists who shop at the farmers' market with cloth bags and make many other small biospheric changes, while still flying tens of thousands of miles per year. Perhaps they simply don't know the enormous impact of their flights (we'll quantify this and other impacts in Chapter 9). This knowledge is a key lever for gradually changing oneself.

Wanting and ego

One Saturday morning, Sharon and I were meditating while the boys were watching *The Empire Strikes Back* in another room. Apparently Darth Vader had just cut off Luke's hand, because we heard Vader say, "Join me, and together, we can rule the Galaxy as father and son!" Out loud, I said, "Why would anyone *want* to rule the Galaxy?" We both laughed. This particular desire seems so ran-dom, so bizarre. Why *would* anyone dream of "ruling the Galaxy?" I honestly wouldn't want to. It sounds like a hassle.

Perhaps longing to "rule" arises from the never-ending cycle of craving and aversion, one of the deepest habits of our lizard brain. People want pleasant things, and they want to avoid un-pleasant situations.[15] When we get something we want, we realize that we still aren't happy. So we want more, and feel that *then* we'll be happy. And of course, when we don't get something we want, we feel miserable. We also want others to have an image of us as perfect and powerful. We suppose (incorrectly) that if we ruled the Galaxy, our every craving would be satisfied. It's difficult to see how a deep habit of wanting traps us in misery—unless we take a careful look within.

Many are the avid environmentalists
who shop at the farmers' market with cloth bags,
while still flying tens of thousands of miles per year.

Our industrial capitalist culture has elevated wanting into an end in itself, the purest of virtues. It's what propels every Disney hero and heroine to live happily ever after. And what is the American Dream if not an attempt to put material desire on a pedestal? We have been led to believe that our desire is what makes us human, and that we can have whatever we wish upon a star.

I believe that *our kindness is what makes us human*, and that our wanting is a huge barrier to kindness. When we give up wanting, we don't become rudderless and unmotivated. Far from it: we become motivated to help others.

Wanting distracts us and agitates us, keeps us out of the present moment, and prevents us from seeing the miracles all around. When we stop wanting, things we took for granted become miracles. Every child is a miracle. So is every animal, every tree, every flower. Every fruit is a miracle, woven from sunlight and earth. That we can eat the fruit and be nourished is a miracle. Sitting in the warm sun on a cool morning is a miracle. So is every breath.

Life isn't drudgery, and neither is it a quest for some grand, mystical, once-in-a-lifetime, transformative miracle. Life is a constant stream of everyday miracles. When we stop wanting, we are happy.

The Great Turning: From wanters to miracle-seers

So, what new story can replace the myths of progress and separation and the primacy of selfish wanting? How can we go to that story? How long will it take for humanity to get there? And what can we do in the meantime?

The new story will be helping each other wake up, and being happy right here, right now. In this new story, instead of chasing progress and growth, we seek to go more deeply into helping one another, living according to our principles, coming out of our prisons of desire, and becoming truly happy.

> Well, Frank, my thoughts are very similar. The vast loneliness up here...makes you realize what you have back there on Earth.
>
> — JIM LOVELL, Apollo 8 telecast from lunar orbit

How can we go to that story? It's possible to come out of wanting, but it's not easy. In Chapter 11, I describe a simple meditation practice for developing awareness of and gradually coming out of the habit of wanting. There are other practices in this book which can support this journey, such as opting out of the consumer economy, but meditation is the most direct path I know. Meditation has allowed me to begin to see how my own wanting works.

How long will it take us to get there? I don't know. I see people around me waking up and helping others to wake up, but it's a long and gradual process; and I see many others who aren't taking even the first step. I have faith that beings living on this planet will get there eventually, but I don't know if it will take 50 years, 50,000 years, or 50,000,000 years.

What can we do in the meantime? Responding to this question is the task of the second part of this book. The short answer is that we can *move ourselves* along the path. After all, what else can we do? My own experience informs me that as we change ourselves for the better, we become more able to change the world for the better—at least our corner of the world. But the main reason to begin walking is that this path is a richly satisfying one, and well worth walking on for this reason alone. Perhaps you're already experiencing this.

We humans *are* capable of coming out of our wanting. We *are* capable of putting others before self, and we can learn how to do this all of the time, not just some of the time. This might be the most important change in our minds that needs to happen. It's the way to come out of the nightmare of fear and violence that we have called history. And while the new story I describe is certainly not a quick fix, it's possible. It's not mystical or abstract. It's not something that just sounds good. It's practical and doable.

This is very clear to me. I hope this book will help you see this, too.

PART II

A MAMMAL IN THE BIOSPHERE

May the outward and inward man be one.

— SOCRATES, from Plato's *Phaedrus*

CHAPTER 7

Trailheads
into the Wilderness

Faith is taking the first step
even though you don't see the whole staircase.

— MARTIN LUTHER KING, JR.

Part I of *Being the Change* attempted to explain why the age of fossil fuels is nearly over. During the last 200 years, fossil fuels transformed the human experience. Today, fossil-fueled consumption is destroying the biosphere and isn't even making us happy. Alienation and depression run rampant through our consumer society.[1] It's time to leave fossil fuels and the money-driven culture with which they intertwine, and to begin living well.

Part II is about what we can actually do—as real people with real limitations—to pioneer this next transformation. I'll describe changes I've made to my daily life in response to our predicament.

Is it an extreme path? I don't think so. I'd take one or two small steps, see how I liked them, and then take a few more. After a few years, these small steps toward living in alignment with the biosphere led gradually to a very different daily life. The changes soon came to feel normal and natural.

My purpose in sharing my own experience is to provide a concrete and lived example—my actual experience of practical change, not wishful thinking, speculation, or futurism—and to show that deep change is possible, not all that difficult, and surprisingly

Doing nothing is one extreme.
Giving up fossil fuels cold turkey would be the other.
I'm walking on a middle path.

satisfying. My changes have demanded no more time and energy than many people devote to their hobbies.

Your path will surely be different from mine. You and I have different priorities, different principles, different experiences, different talents, different interests, different responsibilities, and different resources. Each change you make should be in response to your own reality. Some of my suggestions may work for you; others may not. Be curious: each change you make will lead to a different way of seeing the world and your place in it.

Finally, follow your joy, not your guilt. The changes you make should feel satisfying.

Most of my friends know that burning fossil fuels is a problem. Some of them are also aware that the modern industrial way of life, with its fetishization of separation and material wealth, is not conducive to happiness. But few are changing their lives. They have good intentions, but they are doing nothing.

Doing nothing is one extreme. Giving up fossil fuels cold turkey, if that were even possible, would be the other. I'm walking on a middle path.

Ways to begin (or continue)

Here are a few simple, concrete suggestions to help you get started. They've all been a useful part of my own path, but there are thousands of other similar steps you could take. Whatever you do, approach it with a spirit of curiosity and goodwill.

Ride a bike

Bicycling is happiness. You can easily see this for yourself: dust off your old bike (or borrow one) and go for a ride.

Grow a plant

In the spring, get a few seedlings from somewhere, maybe tomatoes. Perhaps you know an experienced gardener who could give you seedlings and help you get started. Every gardener I know is delighted to help other folks get gardening.

Estimate your CO₂ emissions

This may help if you feel overwhelmed about global warming. Estimate your annual gas and electricity usage from your utility bills, estimate the miles you flew over the last year, and estimate the gallons of gas or diesel you burned. Then look at the "Estimate Your Own Emissions" box in Chapter 9 on page 162, which gives factors to convert these estimates into CO_2 emissions, and make a pie chart.

Repair something

Instead of replacing something broken, or hiring someone to fix it, repair it yourself. Sometimes I feel overwhelmed by a repair, as when I tackle some mysterious problem with my old car. Now I recognize this as a feeling of challenge and growth.

Get rid of some stuff

Give away (or sell) something you haven't used in a long while. This action may help you feel unburdened. All material things are in flux, and feeling possessive is a kind of suffering.

Clean up litter in your neighborhood

Take a walk around your neighborhood with a bag, and pick up the trash you see. If you feel irritated by the litter, or perhaps self-righteous as you pick it up, acknowledge these feelings and come back to enjoying your walk. You've made plenty of garbage yourself over the course of your life: it's great to be able to give back to the community in this small way.

Stop using foam food containers

If you happen to feel the world would be better off without all the foam coffee cups and food containers, stop using them! Do you feel empowered? What else can you change?

Stop using palm oil

If cutting down the world's tropical forests to plant oil palm plantations bothers you, don't use palm oil. This may lead to an interesting and fun examination of your everyday life and its relationship to the biosphere: palm oil is now used in an estimated half of all packaged products for sale at the supermarket.[2]

Go on a meditation retreat

This is a bigger commitment, but if you're ready, it will serve you well. There are many meditation or mindfulness practices you could try. If you are interested in trying the kind of meditation I do, sign up for a 10-day vipassana meditation course.[3] The centers ask for no payment. They run on donations from students who wish to share the benefits they've experienced with others.

Join a community group

Find a group in your community engaged in something important to you, and go to a meeting. If you're concerned about global warming, consider joining (or starting) your local chapter of Citizens' Climate Lobby.[4]

Do something kind for a neighbor

Find some way to do a neighbor a good turn. It doesn't have to be big. Give them some produce from your garden, a jar of homemade jam, or just smile and introduce yourself. Don't expect to become friends. Don't expect *anything* in return.

Plant a fruit tree

Choose a variety that you like to eat and does well in your area, and plant it. It will be helpful to talk to other orchardists in your area. You might be surprised to find what varieties of fruit are grown—or were grown—where you live.

Be vegetarian for a month

If you've ever considered vegetarianism, why wait? Try it for a month. If you eat meat during your vegetarian month, realize there's no reason to feel guilty. If you do feel guilty, observe and accept how you feel.

Spend time in a wild area
Recharge and enjoy.

Biospherism and your partner

If you have a life partner, he or she may not understand the need to stop burning fossil fuel in our daily lives. Even if your partner wants to respond to our predicament as much as you do, he or she may take a different approach.

Be careful not to seek validation for your efforts from your partner. Don't try to convert him or her. This creates strain. (I'm speaking from experience.) Let your enjoyment be your validation.

If you do things you find satisfying and joyful, your partner will start to change. But he or she will do this in his or her own way. You can't control or rush it.

If you start meditating correctly, you will also begin radiating *metta*—compassionate love—to everyone, including your partner. When you begin to deeply appreciate your partner's presence, when you smile and express your love freely, amazing things will happen. If you don't believe me, try it and see for yourself.

CHAPTER 8

Like to Bike

Nothing compares to the simple pleasure of riding a bike.
—JOHN F. KENNEDY

A simple but significant step forward on my path of self-transformation and transcendence of fossil fuels has been to ride a bike. Biking is a wonderful catalyst for mindset change. It's so good, in fact, that today I ride not to be low-energy, but because I love to ride.

Getting back on my bike

During my first year in California, I didn't bike; the idea didn't occur to me. There was still dissonance between my intellectual understanding of global warming and the way I lived day to day. Instead, I rode a large, 35-miles-per-gallon motorcycle the six miles to and from Caltech.

As I became more aware of the interconnectedness of the components of our predicament and my daily actions, burning gasoline increasingly began to feel wrong. Eventually it occurred to me to bike. I owned a solid bicycle—a sturdy old touring machine that had once carried me across 1,000 miles of erstwhile prairie—but it was sitting, neglected, in the back of my parents' garage in suburban Chicago. During a visit one Christmas,[1] I asked the local bike shop for a box and shipped it west.

What a revelation! The first time I biked to Caltech, the feat seemed epic. Somehow a bike made traveling six miles through the suburbs into a grand adventure. I felt unsteady and awkward at first, but after a few days, biking began to feel good—as if I were flying—a feeling I'd taken for granted as a kid. As my body grew stronger and I gained confidence, I cut my commute time in half. The neighborhoods along my route became familiar. I came to rely on biking as my primary exercise.

I started doing as many errands as possible on my bike. I began bike-towing my kindergarten-aged kids to and from school. I carted groceries home in their bike trailer. My bicycle trip radius expanded, and I learned to use public transit to expand it even more. I came to see Altadena and Los Angeles through new eyes. Today, I still enjoy urban exploration by bike, which has a unique magic.[2]

Why I like to bike

Yes, cars are faster over long distances and keep out bad weather. But bikes are better than cars in just about every other way:

- Biking is more fun.
- Biking keeps me fit.
- Biking across town is often faster than driving.
- Biking cures my blues.
- There are no traffic jams on a bike.
- Bike commutes, unlike car commutes, don't leave me frazzled—and won't lead to divorce.[3]
- Biking prevents sickness, at least for me.[4]
- Biking saves serious money.[5]
- Bikes are far easier and cheaper to maintain than cars.[6]
- A good bike can last a lifetime; most cars are junk after eight years.[7]
- Car ownership is a web of obligations and expenses; biking is simplicity and freedom.
- Biking supports local businesses.[8]
- Bikes are easy to park—right next to your destination.
- Biking is the antidote to urban sprawl, and over time will lead to more beautiful neighborhoods.
- Biking is quiet.

- Biking leaves the air clean and doesn't cause respiratory disease.
- Biking is safer for you and for others (more on this below).
- Biking is sexy.
- Biking is adventurous.
- Biking is great thinking time.
- Biking reminds me to enjoy the ride.

Our bodies are made to move!

Biking is actually safer than driving

Let's carefully address the question of whether biking is more dangerous than driving. Biking does carry risk, and should be undertaken with care, but research unequivocally shows that riding a bike is *safer* than driving a car. This truth probably comes as a surprise. It surprised me, at least.

As you'd expect, the risk of injury on a bike is higher. But people who bike regularly are so much less likely to die from heart disease (heart attacks and strokes) that on balance biking is safer. When all causes of death are taken into account, and when averaged over a large enough population, it turns out that bicycling is roughly ten times safer than driving (not bicycling) because the health benefits outweigh the risks.

De Hartog et al. estimated that the physical exercise due to bicycling 4.7–9.3 miles (7.5–15 km) per day[9] adds between 90 and 420 days of life expectancy, whereas the risk of accidents subtracts between 5 and 9 days and the risk of increased inhaled air pollution from motor traffic subtracts between 0.8 and 40 days.[10]

Overall, their best estimate is that the benefits of a modest (three miles each way) bicycle commute outweigh the risks by a factor of nine.[11] Biking is also less likely to kill others, but that's a separate effect.

This result is corroborated by several other studies. To give one example, Rojas-Rueda et al., who studied Barcelona's bike sharing program, estimate the biking benefit (in all-cause mortality) to be well over a factor of ten.[12] On the other hand, I couldn't find a single peer-reviewed study concluding that driving was safer than biking.

In my opinion, this is revolutionary information. Because of it, I now think about biking and driving in a completely new way. The studies also demonstrate the dangerous human tendency to ignore risks from gradual processes such as heart disease (or global warming), and to give too much weight to risks with lower likelihood but higher immediacy.

Finally, it's possible to tip these already great odds dramatically further in your favor by biking smart: ride sober,[13] follow the rules of the road,[14] avoid dangerous situations,[15] wear a helmet,[16] and use lights at night.[17]

Why are there more bikes in Europe?

Once, when I still flew on commercial airplanes, I took a trip to Hannover, Germany, to meet with astrophysics colleagues. What struck me most immediately about Hannover—especially coming from Los Angeles—was its bicycle traffic.

Hannover has an extensive system of segregated bike tracks along city streets. Bicycles are accommodated in the street system as equals to cars and pedestrians, not as an afterthought. The countryside around Hannover has a network of bike paths[18] running along rivers and lakes connecting the small surrounding towns.

And people in Hannover make great use of their bike infrastructure. Students, business people, mothers with kids, elderly folks are all out rolling along together on the segregated tracks. A friend lent me his mom's old bike, so I rode everywhere, including to the gravitational-wave detector in nearby Sarstedt.

In Los Angeles, by contrast, bikes have barely registered in the urban planning consciousness. Bike infrastructure here typically consists of signs suggesting that motorists "share the road," and there are few bicycles on the roads. In the Pasadena urban core, I can ride for miles and I might see two other bicyclists. One Sunday in 2014, during a seven-mile ride from Eagle Rock back to Altadena, I counted the car and bike traffic. I tallied more than 800 cars[19] but *only one other bike.*

In the US, 0.6% of commutes are on bikes[20] versus 25% in the Netherlands.[21] So the average Netherlander gets on a bike about *40 times more often* than the average American. In Amsterdam between 2005 and 2007, residents got onto bikes more often than they got into cars.[22] What accounts for this stark difference between the US and Europe? As a biker who thinks about these kinds of things while happily biking along, I'll suggest three interconnected reasons.

The first is *mindset*. As the nation was settled, space was something to be conquered; land was something to be "improved." This historically adversarial sense of space, along with the emergence of a cultural fetishization of convenience (convenience as progress) in the mid-20th century, intensified a national love affair with cars. Americans equate cars with speed, status, power, and freedom. If we think of bikes at all, we think of them more as recreational toys than as serious transportation. We've lost the ability to imagine a world not dominated by cars.

The second reason is *urban sprawl*, the physical imprint of this mindset. Cities and suburbs in the US tend to be more spread out than in Europe, with vast parking lots for cars and poor integration between residential and business districts. European cities were built before cars existed; their compactness makes them bike-friendly. In Europe, there's simply less distance between where you are and where you need to go.

The third reason is a *lack of bike-friendly infrastructure* in the US, an unsurprising side effect of the car's cultural dominance.[23] Unlike bikers in Europe, bikers in the US must deal with mega-intersections, hostile and distracted motor traffic, and too-narrow roadways with traffic whizzing by, all while managing the risk of being doored by someone getting out of a parked car. Because of this lack of infrastructure, biking here is thought of as too dangerous. In Europe, segregated bike tracks that connect residential neighborhoods to shops, schools, and places of work increase both actual and perceived safety, making biking viable for even timid or elderly riders.

There's evidence that if bike infrastructure improves, more people ride, which leads to still more bike infrastructure and still more riders. This in turn changes the culture and makes biking seem normal and less dangerous, which causes even more people to ride.

There's also a safety in numbers effect: when there are more bikes out, motorists are more aware of bikes. Every doubling of the number of bikers in a community reduces risk of injury from motorists by over 30%.[24] This increases the *real* safety of biking. It also increases the *perceived* safety of biking. Biking encourages more biking. In the long run, a culture of biking could even reverse urban sprawl, because bicyclists tend to advocate for policies that encourage local businesses and compact, integrated neighborhoods.

Climate impact of biking vs. driving

Of course, another major advantage of bikes is that they emit less CO_2 than cars. However, despite what you may think, biking isn't necessarily emissions-free.

Let's compare a mile of driving to a mile of biking. A typical car gets 25 miles per gallon, and burning this gasoline emits 0.5 kg CO_2 over a mile.[25] It also took fossil fuels to manufacture the car; the climate impact embodied in a typical car is between 9 and 20 tonnes CO_2e.[26] Cars typically last 150,000 miles, so embodied emissions add an additional 0.1 kg CO_2e. A typical driven mile therefore emits a total of 0.6 kg CO_2e.

Now consider a typical biker. Riding a bike burns about 50 kilocalories per mile above the resting rate.[27] Producing the food for a typical vegetarian[28] for one year creates 1.5 tonnes CO_2e (see Chapter 9) which comes out to 0.1 kg CO_2e per 50 kilocalories— one vegetarian bicycle mile. A typical meat-eater doubles this, and a proficient organic gardener or freegan could perhaps zero this out. The embodied emissions of the bicycle come out to an insignificant 0.004 kg CO_2 per mile.[29]

On a per-mile basis, then, a 25 mpg car emits six times as much as a vegetarian on a bike. This is a fair comparison for long cross-country trips. But most of our trips are short, in-town trips; and when on a bike, I'm more likely to combine such trips and to choose nearby options like local mom-and-pop businesses. For these trips, I tend to drive about four times as far as I bike to accomplish equivalent tasks. On this basis, the car emits 20 times as much as a vegetarian on a bike, 10 times as much as a meat-eater on a bike, or 500 times as much as a freegan[30] on a bike.

While biking has less impact than driving, it still has impact. Or, to be more precise, our existence has climate impact according to how our food is produced, and biking is like an intensification of our existence. Riding a bike is certainly more biospheric than driving a car—the biosphere made our bodies, and it's great fun to use them—but ultimately riding can only be as biospheric as our food.

It might be surprising that four meat-eaters on a long-distance bike trip would emit less if they shared a hybrid car. But the takeaway message shouldn't be "cars aren't so bad." It should be "our food system is awful!"

Sharon's perspective

Sharon rides 40 or 50 miles per week, most of it on her bike-plus-train commute to the University of California, Irvine. I asked her for her perspective as a woman bike-commuting on urban streets.

"As a woman commuting to work, I need to bring an extra outfit and be willing to mess up my hair.

"I need to carry my bike up and down stairs in the train stations, because the elevators are slow and crowded. It's always hard to find space on the train, especially at rush hour. Some people are grumpy, and some are helpful.

"When I get on my bike, I feel as though I'm putting myself at risk. There's one part of my ride where I say to myself, 'This is the dangerous part.' I have to cross several lanes of traffic coming off the freeway in order to get to the bike lane on the far side. We need some actual bike infrastructure, not just arrows painted on the street.

"People don't bike because everything about our infrastructure is designed for the convenience of cars. Biking is an uphill battle. It's easy to jump in a car, but biking requires planning and maybe even a kind of underground knowledge.

Biking tips

- I have a bike rack with panniers and a milk crate on top (mounted with U-bolts and cable ties). I prefer not wearing a backpack while biking.
- I carry a spare tube, tire levers, and a small pump in my pannier. The day you don't have your flat kit, for whatever reason, will be the day you get a flat five miles from where you need to be in ½ hour.
- Due to the minimal bike infrastructure in Los Angeles, I ride aggressively for my own safety. I take the space I need to be safe while remaining aware of the changing traffic situation. When I make left turns, I act like a car, taking a lane.
- Watch out for dogs. An unleashed dog once ran out directly in front of my wheel and I was down, all in a split second. Luckily, my only injury was a badly bruised hip.
- The worst injury I've experienced in my life so far was on a bike. I was a teenager, riding at night with no lights and not a care in the world on a very dark street, and I rode into a parked car. They rebuilt my nose, and I spent several days in the hospital. Biking is safer than driving, but it isn't risk-free.

"Despite it all, I like to bike. It's fun, it's healthy, and it involves me in the sensual embodied world even more than motorcycling does. It's delightful to go from Altadena to Irvine without using a car."

For those who don't bike yet, I urge you to put down this book and go for a short bike ride. If you don't own a bike, borrow a friend's or take a test ride at the local bike store. I think you'll be happy you did. For the rest of you, bike on!

- Ride smart and minimize that risk. This means, at a minimum, knowing what specific traffic situations are dangerous and avoiding them. Please study bicyclesafe.com[31] carefully—it's well worth the few minutes it will take.
- If you need an electric assist to get up hills, you can buy a kit that uses a small sprocket to drive the chain.
- I respect traffic rules; it's better for bike advocacy. Except for one thing: I don't come to a complete stop at every stop sign. Stop-as-yield is the law in Idaho for good reasons: it makes biking safer and easier, and enhances overall traffic flow.[32]
- Pro tip: When you get to a red light, position your bike over the induction coil embedded in the road to change the signal.
- Don't procrastinate on bike maintenance. Having a bike in perfect working order is wonderful.
- If being on a bike makes you happy, bike with a smile and a wave.

CHAPTER 9

Leaving Fossil Fuel

More of my nights are in a plane than in a bed,
maybe even counting hotels.

— TOM STUKER, first consumer to fly a million miles in a year

So far, by making changes to my daily life, I've reduced my emissions by more than a factor of ten. I used to emit slightly more greenhouse gases than the average American. Now I emit less than the average human (see Figure 9.1). However, I still emit nearly twice the average Bangladeshi,[1] and infinitely more than a wild, nonhuman Earthling.

This level of reduction, while incomplete, allows my family and me to continue a normal suburban life. This suggests that a similar reduction is well within reach for many of us. And as more people make significant reductions, and systemic alternatives to fossil fuel become increasingly available, going the rest of the way will become easier.

To accomplish my reduction, I first had to understand the sources of my climate impact. Basic quantification enabled me to address the biggest sources first. Figure 9.2 shows my estimated emissions of greenhouse gases for 2010. Air travel dominated everything else, but I'd been entirely unaware of this! The simple act of quantification shattered my preconceptions and shined a bright light onto my ignorance.

Quantification allowed me to develop awareness of my daily actions, which was a prerequisite to changing them. Gradually, I've become more aware of what I'm doing in any given moment, and

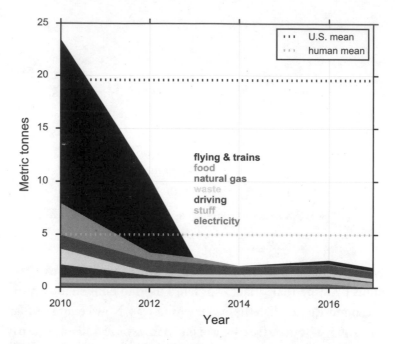

FIGURE 9.1. My personal greenhouse gas emissions from 2010 to 2014, in metric tonnes of CO_2-equivalents (CO_2e), for seven categories. Category labels are in same vertical order as swaths. Horizontal dotted lines indicate average US emissions and average human emissions. The estimates for flying include only CO_2 emissions, and exclude contrail, cirrus, and NO_x effects.

how my actions affect other beings. Daily life is a series of choices; in US society, many of these choices lead to fossil fuels. Finding alternatives requires creativity. For me, it has felt like a fun and engaging game.

Figure 9.2 also shows my 2012 and 2014 emissions. As I ramped down flying, I also reduced my emissions from food, waste, and driving. I became vegetarian, started growing food, and began obtaining much of my food from the waste stream (freeganism). I started composting aggressively, and biking became my primary mode of transportation.

FIGURE 9.2. (*opposite*) US mean per capita emissions (*top*) and my emissions in 2010, 2012, and 2014, for seven categories. As in the previous figure, the estimates for flying exclude non-CO_2 effects.

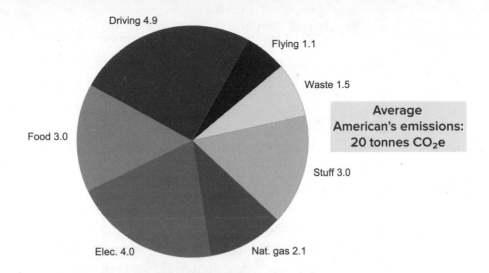

Driving 4.9

Flying 1.1

Waste 1.5

Average American's emissions: 20 tonnes CO_2e

Food 3.0

Stuff 3.0

Elec. 4.0

Nat. gas 2.1

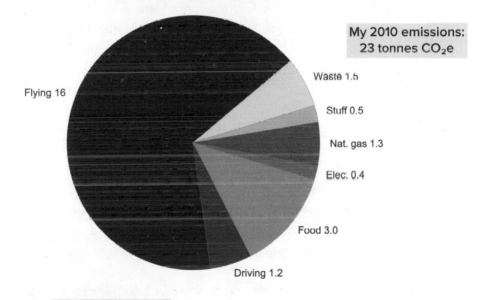

My 2010 emissions: 23 tonnes CO_2e

Flying 16

Waste 1.5

Stuff 0.5

Nat. gas 1.3

Elec. 0.4

Food 3.0

Driving 1.2

My 2012 emissions: 10 tonnes CO_2e

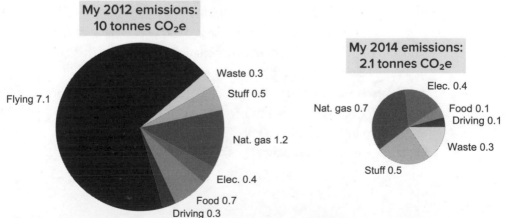

Flying 7.1

Waste 0.3

Stuff 0.5

Nat. gas 1.2

Elec. 0.4

Food 0.7

Driving 0.3

My 2014 emissions: 2.1 tonnes CO_2e

Elec. 0.4

Nat. gas 0.7

Food 0.1

Driving 0.1

Waste 0.3

Stuff 0.5

Why reduce in the first place?

Some people see no value in personal reductions. They haven't cut their own emissions, and they justify their inaction by assuming individual-level reductions don't make any difference. They assume that it requires too much sacrifice. None of this has been true in my experience.

Perhaps surprisingly, my main motivation for reducing was not to keep my own emissions out of the atmosphere. One person's reduction is a tiny drop in a vast ocean of human greenhouse gas emissions. If directly reducing global emissions were my main motivation, I'd find it depressing, like trying to save the world all by myself. Instead, I reduce for three much better reasons.

First, *I enjoy living with less fossil fuel.* I love biking, I love growing food, and I love being at home with my family instead of away at conferences. Less fossil fuel has meant more connection with the land, with food, with family and friends, and with community. If through some magic spell, global warming were to suddenly and completely vanish, I'd continue living with far less fossil fuel.[2]

Second, by moving away from fossil fuel, *I'm aligning my actions with my principles.* Burning fossil fuel with the knowledge of the harm it causes creates cognitive dissonance, which can lead to feelings of guilt, panic, or depression. Others might respond to this cognitive dissonance with cynicism, or perhaps by denying that fossil fuels are harmful. But I find that a better option is simply to align action to principle.

Finally, I believe *personal reduction does help, indirectly, by shifting the culture.* I've had countless discussions about the changes I've made, and I've seen many people around me begin to make similar changes in their own lives. By changing ourselves, we help others envision change. We gradually shift cultural norms.

The essential story I'm telling is that life can be better without fossil fuel. I've experienced this to be true; if others also experience this to be true, who's to say the story won't develop a powerful, change-making resonance?

Of course, there's no reason to limit our actions to the personal sphere. I find that actions aimed at the personal and actions aimed at the collective are mutually reinforcing. I see no reason not to do both.

Making the estimates

This section gives the details behind my estimates as well as the tools you need to assess your own situation. But first, a few general considerations.

> It is remarkable how long men will believe in the bottomlessness of a pond without taking the trouble to sound it.
>
> — HENRY DAVID THOREAU, *Walden*

Your emissions profile will not look like mine

The average US citizen's emissions profile (Figure 9.2) looks nothing like my 2010 profile. Your emissions profile is likely also unique. Your life is different from mine, and you must understand the idiosyncrasies of your own situation before you can take meaningful action.

Some actions are bigger than others

These days, there are lots of suggestions for how we can "save the planet," but very little about their relative effectiveness. This lack of distinction paves the way for greenwash (for example, airports "going green"). Lack of quantification also kept me from realizing the impact of my flying, despite my deep concern over global warming.

Not everyone's emissions will be as dominated by flying as mine were. Many US citizens don't fly at all, so national average emissions from flying are low. Someone close to this average profile—whose major emissions are from driving, new stuff, electricity, and food—should naturally focus on things like bicycling, living close to work, buying less stuff, sourcing renewable electricity, and eating less meat.

I've represented estimates of the relative reductions from my major actions in Figure 9.3. My five most effective actions were

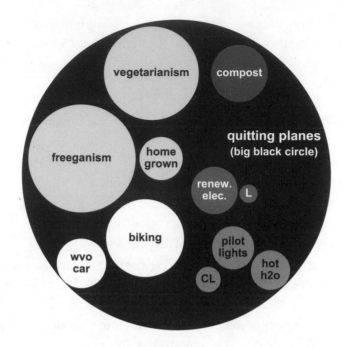

FIGURE 9.3. My reductions proportional to areas of circles. Quitting planes is the all-encompassing black circle. L indicates taking care to turn off lights; CL indicates switching to a clothesline.

quitting planes, vegetarianism, bicycling, freeganism, and composting. This is such an important point—that not all actions are equal in terms of emissions reductions—that I've also tabulated the estimates in Table 9.1.

Physics bats last

Our predicament consists of many interconnected processes, systems, and mindsets. It's larger than global warming. Being locked into the old story of separation and amplification of our egos (by reacting automatically with craving and aversion) underlies much of this. Therefore, anything we do to counter these bad mental habits and to build a more beautiful world helps at some level.[3] For example, the moment in which we observe an impulse to say something negative—and we stop ourselves—is a wonderful

TABLE 9.1. Emissions reductions from some of the changes I made to my daily life. Note that your potential for reductions from these changes depends on your own particular situation: if you already fly very little, you won't see a 16.5 tonne reduction from quitting planes! Also, you might experience much larger reductions than I did by switching to renewable electricity and solar hot water.

Action	Annual reduction
quitting planes	16,500 kg CO_2
freeganism	2,000 kg CO_2e
vegetarianism	1,500 kg CO_2e
bicycling	1,000 kg CO_2
composting	700 kg CO_2e
driving on WVO	400 kg CO_2
growing food	300 kg CO_2e
reducing pilot lights	300 kg CO_2e
renewable electricity	300 kg CO_2e
solar hot water	200 kg CO_2e
humanure	150 kg CO_2e
clothesline	100 kg CO_2e
turning off lights	50 kg CO_2e

moment. So is the moment in which we help another without expecting anything in return

However, the fact remains that humanity urgently needs to stop burning fossil fuels. Photons interacting with the atmosphere don't respond to our acts of kindness; they only respond to our acts of greenhouse gas emission. An appropriate response to our predicament, then, must also include actions that target emissions. These concrete, focused actions, taken by enough individuals to create cultural shift, will begin to change the systems and institutions that lead us to mindlessly burn huge amounts of fossil fuels in our daily lives. One of the most effective actions any of us can take as an individual is to radically reduce our own fossil fuel use. Doing this even caused my advocacy for collective action to become more resonant and powerful.

TABLE 9.2. Climate impact conversion factors, as discussed in the text. I've shown CO_2-only emissions for curiosity's sake; for quantifying your emissions, I recommend that you use the CO_2e values.

	Unit	kg CO_2e	kg CO_2
airplane (coach)	per passenger mile	0.8	0.3
airplane (1st class)	per passenger mile	1.5	0.6
car (25-mpg, no passenger)	per mile	0.452	0.452
Amtrak train (coach)	per passenger mile	0.14	0.14
bus (long haul)	per passenger mile	0.065	0.065
gas	per gallon	11.3	11.3
diesel/fuel oil	per gallon	12.1	12.1
natural gas	per therm	13	6.1
food (mean American)	per person year	3,000	
food (vegetarian)	per person year	1,500	
food (vegan)	per person year	1,000	
municipal sewage	per person year	150	
municipal waste	per person year	1,300	
electricity (US)	per kWh	0.9	
new stuff	per US dollar	0.5	

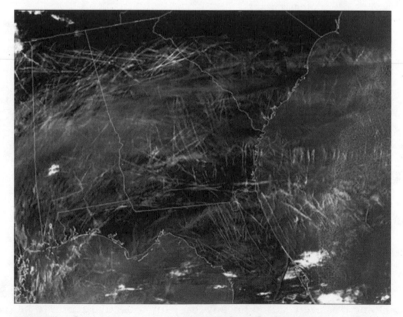

FIGURE 9.4. Contrails over southeastern US on October 13, 2004. High ice clouds like contrails and cirrus are mostly transparent to incoming solar (shortwave) radiation, but mostly opaque to outgoing thermal (longwave) radiation; they therefore warm the climate.

Let's dive in

The yardstick we'll use for measuring climate impact is kilograms (kg) or metric tonnes (1 tonne = 1,000 kg) of CO_2-equivalents (CO_2e). As discussed in Chapter 3, climate impact from other sources (such as methane) is converted into CO_2e, a greenhouse-effect-equivalent amount of CO_2, for apples-to-apples comparison.

Table 9.2 gives conversion factors from daily life units to CO_2e amounts.[4] (I'll explain how I came up with them below.) These estimates include the "upstream" emissions generated from obtaining and processing the fuels, which add an additional 20% or so[5] beyond what's emitted by burning the end product. Note that upstream emissions are rising as we increasingly turn to energy-intensive production methods like fracking, deep offshore drilling, and tar sands extraction.

Air travel

Hour for hour, the quickest way a normal human can warm the planet is by flying on a commercial plane. Airplanes contribute to global warming via CO_2 emissions, NO_x emissions,[6] contrails, and cirrus cloud formation (Figure 9.4). High icy clouds such as contrails and cirrus let the sun through but trap outgoing thermal radiation from the ground, warming the planet.

The global warming contribution from the average plane's CO_2 emissions alone is 0.3 kg CO_2 per passenger mile.[7] That's for coach; first-class or business-class fliers are responsible for twice these rates of emissions, as they take up over twice the space on a plane.[8] Then, the non-CO_2 effects likely double or triple the global warming potential of airplanes.[9] Unlike CO_2, these effects are short-term: if planes suddenly stopped flying, they'd disappear within days.

The act of flying is an exercise of privilege. Globally, only about 5% of humans have ever flown.[10] The average American emits about 1,000 kg CO_2[11] per year from flying,[12] which is roughly equivalent to one 4,000-mile round-trip between Los Angeles and Chicago. In 2010, however, I flew 50,000 miles, equivalent to two transatlantic and six domestic flights. My flights that year resulted in 16 metric tonnes of CO_2.

Academics are expected to fly to universities and conferences to talk about their research. This provides visibility for job offers and funding. Many of my friends and colleagues fly more than 100,000 miles per year; climate scientists fly just as much as astrophysicists (and other academics). There's little interest in teleconferencing.

Nonetheless, flying in commercial planes began to feel increasingly unwholesome to me.[13] I gradually ramped down my air travel as it dawned on me that the benefits to my career weren't worth the negative consequences.[14] Who would pay the real cost of my flights? I decided I couldn't justify putting my caree. _.__ the biosphere. I imagined my kids someday saying, "Dad, how could you keep flying around the world giving 20-minute updates on your research when you knew the impact?" They would be right to ask this.

I now prefer slow and adventurous travel to fast travel. By not flying all the time, I'm able to connect much more deeply to my local community, which is satisfying. Nor do I miss the jet-setting life: the hotels, the homesickness, the airport lines, the jet lag, and the colds I'd often catch on planes. (Indeed, modern air travel is the ideal way for a local outbreak to spread into a pandemic.)

I decided to only include the long-term CO_2 portion of airplane impact in the above pie charts, which makes airplanes look much better than they really are. But either way, the basic conclusion was inescapable: reducing my emissions would require flying less.

Car travel

My largest climate impact was from flying, but the average American's largest climate impact is from driving. In the US, the average person emits about 5,000 kg CO_2 per year from driving.[15] A map of any city's CO_2 emissions looks exactly like a map of that city's expressways (or as Sharon calls them, petrol deathways[16]).

Here we're only considering the impact from fuel—burning a gallon of gas emits 11.3 kg CO_2, while burning a gallon of diesel emits 12.1 kg CO_2[17]—but keep in mind that driving requires roads, bridges, parking, and the cars and trucks themselves, the creation of which also produces emissions.

In 2010, Sharon and I burned 330 gallons of gas to drive a 45 mpg gasoline car about 15,000 miles. I'll claim ⅓ of these emissions (1,200 kg CO_2), a guess, attributing the rest to Sharon and our sons.

Since 2011, most of my driving has been in Maeby, a 1984 diesel Mercedes-Benz fueled by waste vegetable oil (WVO). Burning plant oil doesn't increase the net CO_2 in the atmosphere. The plants capture CO_2 from the atmosphere to make the oil, and burning the oil releases that same CO_2 back to the atmosphere for no net gain.[18]

Maeby is useful for hauling heavy or bulky materials or for traveling great distances. However, for trips of fewer than ten miles, I prefer riding my bike. Biking is more fun, which is saying something, as I still get a thrill when I flip the two switches on my dashboard that take Maeby from diesel to veg.

Maeby requires a cup of diesel to warm up before switching to WVO. In 2012, my personal use of diesel fuel in Maeby came out to 13 gallons (I kept a log for three months, then multiplied by four). In addition, I burned about 10 gallons of gasoline in other cars for a total of 300 kg CO_2. In 2013, my portion was three gallons of diesel and two gallons of gas, for total estimated emissions of 50 kg CO_2. The improvement came from more biking and less driving.

Switching from fossil fuel to WVO eliminated 400 kg CO_2 per year. But biking reduced my annual emissions by 1,000 kg CO_2 as it became my primary mode of transportation.[19]

Methane, also known as "natural gas"

Natural gas is often touted as a bridge fuel, a clean pathway for transitioning from fossil fuels to renewables. This is because natural gas emits only half as much CO_2 as coal on an energy-equivalent basis.

However, natural gas also leaks into the air at every step in its life cycle, and it's a far more potent greenhouse gas than CO_2. As gas leaks from wells, pipelines, processing plants, and the 100-year-old cast-iron pipes under our streets,[20] it contributes significantly to global warming.

After refining, natural gas is essentially 100% methane,[21] and the global warming potential (GWP) of methane is about 105 and 33 times that of CO_2 on 20-year and 100-year time horizons, respectively.[22] Here, I'll use the average value of the 20- and 100-year time horizons, assigning 1 kg of methane a GWP of 65 kg CO_2e.[23]

> The gas extracted from shale deposits is not a "bridge" to a renewable energy future—it's a gangplank.
>
> — Anthony Ingraffea, fracturing engineer

At a GWP of 65, methane leakage of 4% or more would make natural gas worse than coal in terms of climate change.[24] A 2011 study by Howarth et al. estimated total leakage at between 3.6% and 7.9% for gas produced by fracking and 1.7% and 6% for conventional gas.[25] Subsequent airborne measurements over Uintah County, Utah, estimated leakage at between 6.2% and 11.7% of production.[26] Because of this leakage, an energy system built on natural gas might well be worse for the climate than one built on coal.

Meanwhile, energy utilities and municipalities in the US are scrambling to transition to natural gas, because they mistakenly view it as the "green fossil fuel." In terms of global warming, this is a mistake.

A major regulatory overhaul of natural gas drilling operations combined with a major infrastructure overhaul could eliminate some of the leaks. However, to be meaningful, this overhaul would need to include the old pipes running under our streets, and this would be extremely expensive: laid end to end, the natural gas pipe in the US alone would stretch to and from the moon almost three times.[27] Perhaps we'd be wiser to leave natural gas and gas appliances behind as quickly as possible, replacing them with carbon-free electricity instead.

Burning a therm of natural gas emits 5.3 kg CO_2; including the upstream emissions bumps this amount to 6.1 kg CO_2. To account for leakage, and since half of the gas consumed in the US is fracked, we can use the mean of the Howarth et al. ranges (5% methane leakage). This translates into an additional 6.5 kg CO_2e from methane leakage per therm of natural gas,[28] for total emissions of 13 kg CO_2e per therm. The average American emits 2,000 kg CO_2e per year from residential natural gas.[29]

When we first moved into our house in 2008, there were five pilot lights emitting a whopping 1,600 kg CO_2e per year. We shouldn't still be using pilot lights in an age of global warming! Eliminating two stove pilot lights cut 600 kg CO_2e, and turning down the rest (oven, space heater, water heater) cut an additional 300 kg CO_2.

In 2012, we used 372 therms of natural gas—4,400 kg CO_2e— for heat, cooking, hot water, and drying clothes. We then switched to using a clothesline, which reduced our household CO_2e emissions by 550 kg.

In 2016, our old gas water heater died, and I decided to replace it with a solar hot water heater. This reduced our household emissions by about 800 kg CO_2e per year (most households would see a larger reduction[30]), but it was expensive at $4,000.[31] I could go further down this route, replacing our old gas heater with an electric heat pump for around $5,000, and replacing our gas range and oven with an electric induction range and convection oven for $2,000.

These electrical devices would require upgrading our home's outdated electrical wiring and service, which I expect would cost more than $7,000. For a total of $18,000, we could eliminate 2,500 kg CO_2e in natural gas emissions per year. All this new stuff would have roughly 12,000 kg CO_2e in embodied emissions (see below, and realize that the water heater had a rebate), so the emissions payback time, if running on renewable electricity, would be five years—not terrible, but not great, either. But we can't afford all this. And for us, it wouldn't even make sense from an annual operating cost perspective.

Food

My second-largest climate impact came from food. This was another big surprise, revealed to me only by looking at the numbers.

Growing, processing, packaging, and distributing food produces greenhouse gases: CO_2 from fuels, fertilizer production and land-use changes such as deforestation; nitrous oxide (N_2O) from fertilizer production and application;[32] and methane from livestock. About ⅓ of global greenhouse gas emissions are due to food production,[33] and about ½ of this (15% of global emissions) is due to livestock,[34] mainly cows.

Your individual emissions from food production naturally depend on your diet—what you eat, how much you eat, and how your food gets to your plate. Producing the food for typical meat, vegetarian, and vegan diets emits about 3,000, 1,500, and 1,000 kg CO_2e per year, respectively.[35] The average American diet emits 2,900 kg CO_2e per year, slightly less than the average meat diet, since 3% of Americans are vegetarian (about half of whom are vegan).[36]

In 2012, I stopped eating meat primarily to avoid harming animals, and I personally prefer vegetarianism. In addition, it reduced my emissions by about 1,500 kg CO_2e per year. Over the next few years, I began growing food, trading surpluses with neighbors, and rescuing supermarket discards (freeganism). Most of my food now comes from these sources. I estimate that freeganism alone reduces my food emissions by an additional 1,000 kg CO_2e per year.[37]

Electricity

You wouldn't dream of leaving your water taps wide open when you're not using them; I feel the same way about electricity. Electricity is precious.

Emissions from electricity depend on the method of generation. If your electricity is sourced from 100% renewables, then you have no electricity emissions. Otherwise, generating electricity in the US emits 0.9 kg CO_2e per kWh on average.[38]

The typical American consumes 4,300 kWh of residential electricity per year, emitting 4,000 kg CO_2e.[39] In our house, we somehow use less than ¹⁄₁₀ of this per person. We don't have a TV or air conditioner, and we're careful to turn things off, but otherwise we use electricity liberally. I suspect that much residential electricity in the US is simply wasted, and that it would be quite possible to cut our national consumption by ½ or more without real sacrifice.

Rooftop solar probably makes sense for most 11,000 kWh-per-year households, but at our annual usage of only 1,200 kWh, it's not yet cost-effective. Leasing isn't an option for such a small system. But in 2016, SoCal Edison finally offered us a 100% renewable option. This will cost our household an additional $50 per year,[40] a far better deal for us than rooftop solar.[41] Before this option, my share of our household emissions came to 300 kg CO_2e per year; afterwards, of course, it dropped to zero. I suspect SCE is responding to growing demand for community choice aggregation, an exciting strategy for switching entire regions to 100% renewable electricity, building community, and raising climate awareness (see Chapter 15).

Stuff

Manufacturing new stuff—cars, clothes, computers, furniture, and so on—emits greenhouse gases. Estimating these emissions rigorously would be quite complicated, but for our purposes, a simple rule of thumb will suffice: each dollar spent on new stuff represents roughly 0.5 kg embodied CO_2e emissions (counting manufacturing, packaging, and shipping).[42]

The average US person spends a little over $6,000 per year on new stuff.[43] (Remarkably, ⅓ of this goes toward new cars,[44] another measure of the automobile's dominance in our lives.) Therefore, average emissions are something like 3,000 kg CO_2e. My wife and I both prefer not having much stuff; our four-person household spends about $4,000 per year on goods (clothes, books, hardware, stuff from Target, etc.) My portion of these emissions is 500 kg CO_2e.

These estimates don't include housing. Building a modest two-bedroom house with modern industrial methods produces some 80 tonnes CO_2e.[45] We could divide emissions from long-lived items by the item's lifetime to get annualized emissions. Our bungalow was built in 1926, and if it lasts another 60 years, we could assign 100 kg per year for each of the four people in my family. However, distributing the emissions in this simplistic way is clearly not right, because it doesn't distinguish between buying new versus used. A careful accounting of such long-lived purchases should assign more emissions to the earliest years.

Waste

When we send our stuff to the landfill,[46] the organics—food waste, yard waste, paper, and textiles—decompose anaerobically (without oxygen), producing methane. US landfills emit 1,300 kg CO_2e per person per year.[47] Let this sink in for a moment: our society has reached a point where even one person's trash, taken by itself, generates more CO_2e than the average Bangladeshi generates for everything.[48]

Sewage treatment plants in the US generate an additional 150 kg CO_2e per person annually, due to the anaerobic decomposition of your "personal waste," along with waste from industrial meat production; and also due to the electricity required to run the industrial-scale processing plants.[49]

This analysis highlights the need for change at the systems level: as a society, we should never, ever put organic material into landfills. It also highlights the interconnection between our food and waste systems, which I've begun thinking of as one integrated system. For example, two-thirds of food waste occur before the supermarket checkout line; as an individual, you have no control over that methane-emitting waste except by opting out of the industrial food system (see Chapter 13).

Still, composting is a wonderful practice. If you have a yard, you can safely compost your food waste, yard waste, and human waste (Chapter 12). This eliminates some of these waste emissions since backyard piles decompose aerobically, emitting CO_2 instead of methane. Humanure cuts up to 150 kg CO_2e of sewage treatment emissions, and composting food scraps and yard waste cuts another 400 kg CO_2e.[50]

However, this means that, even if you compost aggressively, your landfill waste is still emitting 900 kg CO_2e. On average in the US, ⅓ of this comes from the pre-consumer food waste I've already mentioned, ⅓ comes from non-recycled paper, and ⅓ comes from textiles.[51] You can recycle nearly all your personal paper waste. What about clothes? I've tried composting rags and old jeans, but it didn't work: the cloth decomposed too slowly. Until we design clothes for the home compost bin, we can find ways to opt out of buying new cloth, for example by exchanging used clothes with one another.

Buses, trains, and a transportation comparison

Riding Amtrak in coach class emits 0.14 kg CO_2 per passenger mile.[52] Long-haul buses (in this case, Greyhound) have about half the emissions of Amtrak: 0.07 kg CO_2 per passenger mile.[53]

How the different travel options stack up is illustrated in Figure 9.5. For a given distance, carpooling four to a 50 mpg car emits less than taking the train, which emits less than a 50 mpg car (alone), which emits less than flying. Four people flying on a plane (in coach) emit 5 times the CO_2 (and 14 times the short-term CO_2e) as four people sharing a 50 mpg car. Keep in mind that we typically go many more miles in a plane than in a car, bus, or train. The average Amtrak ride, for example, is 218 miles.[54]

Trains still have potential for efficiency increases, and could run on renewable electricity. They therefore hold potential to become much more favorable relative to other modes of transport. But planes are already nearly as efficient as they can be.[55]

Everything else

The categories we've considered so far—flying, driving, food, residential electricity, natural gas, stuff, and waste—come to a total of 20 tonnes per year for the average American. But what about emissions sources outside of these categories, such as roads, buildings,

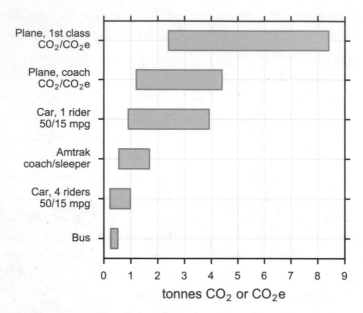

FIGURE 9.5. Personal emissions for 4,000 miles of travel, equivalent to a round trip between Los Angeles and Chicago. For planes, the left and right sides of the bars give emissions in CO_2 units and CO_2e units (including short-term non-CO_2 effects), respectively. For cars, the left and right sides of the bars give emissions for 50-mpg and 15-mpg cars, respectively.

hospitals, and non-residential electricity? If we know the total emissions, we can get a rough idea about these other emissions.

According to the US EPA, in 2013 the US emitted a grand total of 7,700 million tonnes of greenhouse gases.[56] Dividing by the US population, we find annual *per capita* emissions of 24 tonnes, 4 tonnes in excess of the total from our seven categories. This suggests that these other emissions amount to around 4 tonnes per person per year.

Low-energy living saves money

Changing to a low-energy lifestyle has saved my family thousands of dollars. We spend less than one-half what we used to on food, we buy far less gasoline and diesel, we spend less than $20 per month on electricity, and we save on air travel. We also see indirect savings, for example, from staying healthier.

By contrast, many of my friends have reduced their emissions by buying electric vehicles (EV), rooftop solar panels (PV), and other expensive equipment. While still out of reach for many, EVs are becoming more affordable. And I know folks who have saved $30 or $40 per month over their former monthly electric bills of $100 or more by "going solar." Also, in my opinion, the embodied emissions payback times are impressively short. A new electric car represents seven tonnes of embodied CO_2e emissions, equivalent to burning 700 gallons of gasoline,[58] for an average payback time of about 1.5 years.[58] Solar panels require three or four years to pay back their manufacturing emissions.[60]

The shopping option allows people to reduce their carbon emissions without changing their lifestyles. This is good in that it helps people get started easily. However, it comes from a consumer mentality which may be a barrier to going further. Instead of examining behaviors and changing themselves, shoppers reinforce a comfortable habit of consuming. Isn't this a big part of why we're in such a mess to begin with?[61] The shopping option might be a step in the right direction, but it means even more stuff; if it changes mindsets, shopping does so only incrementally; and it's out of reach for the vast majority of humanity.

I chose a no-cost, home-brewed path because I enjoy changing my lifestyle, spending less, and opting out of consumerism. Part of my experiment in low-energy living is to show that anyone can do it, regardless of income level.

Flying and the quest for sustainability

I know many people who strive to live sustainable lives. They eat raw vegan diets, use hand-crank juicers, and keep goats for milk and meat.

Yet the raw vegan flies from the US to Asia several times per year, and the goat keeper flies across the continent once per month for her business. As we've seen, their flights cast a long, unsustainable shadow over their other efforts. Why do many people who strive to be "green" nonetheless choose to fly frequently?

Part of the answer, I think, is that few people quantify the carbon impacts of their actions, so they don't realize that their flying habit dominates their emissions. I certainly had no idea until I estimated it myself.

Estimate your own emissions

Here's the recipe for connecting your daily actions to your emissions:[57]

- **Planes:** Multiply your annual miles flown in coach and first class by 0.8 kg CO_2e and 1.5 kg CO_2e, respectively, and add the resulting two numbers. (If you wish to exclude cloud, contrail, and ozone warming effects, use 0.3 kg CO_2 per coach mile and 0.6 kg CO_2 per first class mile, respectively.)
- **Driving:** For each vehicle you drove, estimate the miles driven and divide by the vehicle's miles per gallon (mpg); this gives gallons of fuel burned. Then multiply by 11.3 kg CO_2 per gallon of gas or 12.1 kg CO_2 per gallon of diesel. Finally, divide by the average number of people in the vehicle.
- **Intercity trains and buses:** Multiply the miles you rode the bus in a year by 0.065 kg CO_2 per mile. Multiply the miles you rode Amtrak in a year by 0.16 kg CO_2 per mile.
- **Natural gas:** Find your household total annual gas usage from your utility bills (one therm is equivalent to 98 cubic feet or 2.8 cubic meters). Divide this by the number of people in your household, and multiply by 12 kg CO_2e per therm.
- **Food:** Meat diet: 3,000 kg CO_2e per year; vegetarian

However, I also think that many people are addicted to flying. Even if they knew the impact of their choice to fly, I suspect many would continue doing so. When the goat keeper asked me why I don't fly, I told her how flying used to completely dominate my emissions. She said, "Well, I have to fly for my job." The raw vegan told me that buying offsets for her flying is the best she can do "while still fulfilling my adventurous, traveling life purpose." Humans are good at rationalizing petroculture.

diet: 1,500 kg CO_2e per year; vegan diet: 1,000 kg CO_2e. If you get some fraction from your garden or from freeganism, reduce proportionally.

- **Electricity:** Add up your annual household electricity usage in kWh (from your utility bills), divide by the number of people in your household, and multiply by 0.9 kg CO_2e per kWh.
- **Waste:** 700 kg CO_2e from food and yard waste (less 200 kg CO_2e if you compost all food waste, less 200 kg CO_2e if you compost all yard waste, less 300 kg CO_2e if you get at least one-quarter of your food from freeganism); 150 kg CO_2e from sewage (unless you practice humanure). 300 kg CO_2e comes from paper waste (assuming you recycle paper and cardboard, otherwise it would be higher), and 300 kg CO_2e come from textiles. If much of your clothes are secondhand, reduce your textile waste proportionally.
- **New stuff:** Estimate your annual spending on new goods and multiply by 0.5 kg CO_2e per US dollar.

Knowing your emissions across these categorizes, and tracking them over time, will help you make informed decisions as you reduce your emissions.

Climate impact of everyday actions

- **School drop-offs:** The majority of kids in the US arrive by private vehicle, and fewer than 15% walk or bicycle.[62] A family living 6 miles from school drives 24 miles per day for drop-offs, or 4,300 miles for a 180-day school year.[63] If they drive a typical (25 miles per gallon[64]) car, they burn 170 gallons of gas. This emits 1,950 kg CO_2.
- **Heating a house in Los Angeles:** Our heater burns about 100 therms per year to heat our 1,400-square-foot house, about ⅓ our annual natural gas consumption. This emits 1,300 kg CO_2e per year.
- **Drying laundry:** One load emits 3.5 kg CO_2e (including gas and electricity). A household of four running three loads per week creates annual emissions of 550 kg CO_2e.

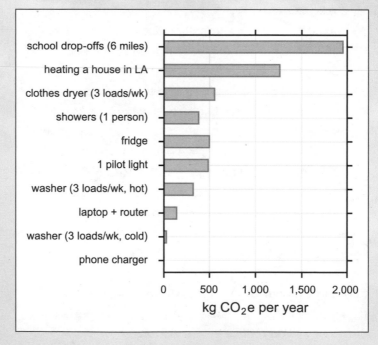

- **Showering:** An average shower in the US uses 17 gallons of water.[65] To heat that water from 50°F to 107°F requires burning 0.083 therms of natural gas (8,300 Btu). In a year, this emits 380 kg CO_2e.
- **Refrigerator:** My fridge uses 1.5 kWh per day (as measured with a widely available, twenty-dollar electric power meter), emitting 1.4 kg CO_2e, or 500 kg CO_2e per year. (Imagine all the refrigerators running over the world each emitting 1.4 kg CO_2e per day.)
- **Pilot light:** The two always-on pilot lights in our house (oven and water heater) burn 77 therms of gas in a year, emitting 970 kg CO_2e![66] An average pilot light emits a whopping 490 kg CO_2e per year.
- **Laundry:** One load takes 0.2 kWh of electricity. Three loads per week on cold adds up to 30 kg CO_2e per year. If we want hot water, we need to burn about 0.15 therms of natural gas to heat the 30 gallons.[67] Over one year, this adds up to 300 kg CO_2e. Heating the water causes 30 times as much impact as the electricity.
- **Computer:** I measured a MacBook Air with the power meter; it uses 34 W. I use a laptop about 8 hours per day; this comes to 100 kWh per year. Our home wireless modem uses 6 W continuously (53 kWh per year). Assuming dirty electricity, the total is 140 kg CO_2e per year.
- **Cell phone charger:** With my phone charger plugged in, the power meter reads 0.0 W. This comes to less than 0.4 kg CO_2e over the course of a year, possibly much less—insignificant. Let's go after things that actually matter, like flying in planes.

Once, after a presentation I gave, a woman came up to talk to me, distraught. She told me that while she was very concerned about global warming, she was also a nurse who flew to Africa several times a year to assist in childbirths. She was saving the lives of mothers and children, and this work was deeply meaningful to her—so naturally she felt conflicted about her flying emissions.

The plane would have flown anyway. Right?

Often, I hear smart people attempt to rationalize their flying by claiming that the plane would have flown without them, so they're responsible for little or none of the plane's emissions. This is a fallacy. Instead, there are at least two logical ways to look at the question.

1. We have a group of passengers all flying on the same plane (for simplicity, let's assume all the seats are identical coach seats). The plane generates some measurable and known quantity of emissions to make the trip. Each passenger chose to take the trip, bought a ticket, and sat on the plane. The passengers' intentions and actions were equivalent, so we'd best assign the same emissions to each, dividing the plane's emissions by the number of passengers.

2. The airline has a business algorithm which schedules planes, and (hypothetically) one of the passengers on the plane critically tipped the algorithm to schedule the flight. Some would argue that this passenger should be assigned most of the emissions from the flight, while the others should be assigned only the emissions needed to fly one additional passenger. But the critical passenger remains unknown, so our best option is to take the expected value, which is equivalent to dividing the plane's emissions by the number of passengers.[68]

I think the answer to situations like these is simple, but also difficult to hear: don't try to live in two places at once. If you have truly meaningful work overseas, dive into its core by living there. Chances are it will then become meaningful beyond your wildest dreams. As my friend Pancho Ramos-Stierle says, "If you want to serve people, you come and live with the people, and laugh with them, and cry with them, and eat with them." And if you can't bring yourself to make this dive, then serve the people in your neighborhood. There are plenty of people to serve within biking distance. It only takes a shift in your thinking to find them. We don't need to get on airplanes to serve.

High-profile environmental leaders also tend to fly a lot. They feel that their efforts to solve global warming justify their huge footprints, and perhaps they're right. Still, I can't help wondering whether they'd be even more effective than they already are, if only they'd walk their talk and lead by example.

In *Being the Change*, I attempt to present facts as accurately as I can and, in so doing, to raise awareness a little bit. Awareness is the soil out of which change grows. In Chapter 10, I describe some slow-travel alternatives that replaced my own flying habit. It took me about two years to change; I, too, was once addicted to flying.

If you choose flying, go forth and fly. But please don't pretend your life is sustainable.

Why carbon offsets don't work

The idea behind carbon offsets is to pay some company a modest sum (around $30 for a transatlantic flight) to "offset" the carbon emission of the flight by planting trees.[69] Unfortunately, this simply doesn't help.

First, carbon offset companies, whether well-intentioned or not, can't guarantee that the trees they plant (assuming they plant any) will be allowed to grow undisturbed for centuries, or that a compensating mass of trees won't be cut down somewhere else on the planet. How could they? They can't even guarantee they'll be in business in two years.

But suppose the offset companies could somehow magically guarantee they'd protect the trees for the next 1,000 years, even in the face of massive development pressure from our exponentially growing population. Even so, the scheme *still* wouldn't work.

As mentioned in Chapter 3, a bit less than ⅓ of our cumulative emissions from 1750 to 2011 were from deforestation, while ⅔ were from burning fossil fuels. Suppose we planted so many trees that we reforested the planet to 1750 levels. (It's hard to see how we could—even a mere 7.5 billion hungry humans *destroy* forests at a rate of 90,000 acres per day,[70] whereas world population is forecast to grow to 11 billion by 2100. But suppose we did.)

By definition, this reforestation would recapture the cumulative emissions from deforestation. But *fossil fuel emissions* would remain.[71] To draw those down as well, we'd need to keep planting trees—three times what we'd need to plant to get back to the 1750 reforestation, which was already implausible.

To look at it another way, all of the trees we could possibly plant are already spoken for by the carbon we've *already* emitted. Suppose we've already emitted 100 units of carbon (the numbers here are meant to be illustrative), and that we now stop emitting and reforest as much as we can. Suppose this best-case reforestation captures 30 units. Afterwards, 70 units remain at large.

What if, instead of stopping immediately, we all took a few more flights and bought offsets? Now we've emitted, say, 101 units. The offsetting company plants some trees. But these trees would have been planted anyway in the best-case reforestation, which still captures only 30 units. But now 71 units remain at large instead of 70.

If buying carbon offsets locked fossil fuel emissions permanently back underground from whence they came, I'd be all for it. But they don't.

People often argue that while offsets aren't perfect, at least they're doing some good. But are they? Offsets allow people to assuage their guilt and continue flying. In other words, offsets may ultimately increase CO_2 emissions.[72] We should certainly plant all the forests we can plant, but this should never be an excuse to continue burning fossil fuels.

Accepting my intertwined-ness

Despite my efforts to transform my daily actions, I'm still involved in the project of industrial civilization. I use roads, I shop at supermarkets, and when my children are injured, I'm grateful for the emergency room. I also had a few decades of emitting much more than I do now.

Nonetheless, two facts stand out from my ongoing experiment in low-energy living: I was able to make changes to my life that cut my global warming impact by well over a factor of ten, and these changes made me happier. From these facts, I draw the following two conclusions.

First: if everyone—especially those of us who live in affluent societies made a conscious effort to move toward low-energy living and to change the collective mindset (and the collective infrastructure), we could quickly and painlessly cut global emissions in half, and with that momentum, we'd probably go much deeper. I'm a busy suburbanite with a demanding job, two small kids, and a wife who also works; the fact that I could easily make such deep emissions cuts means that you could likely do so, as well.[13] If we all changed together, I'm certain that we'd amaze ourselves with our collective power.

Second: consuming less, and emitting less, makes me happier. Realizing this truth requires a shift in mindset, but it's a shift more and more people are making. My own experiments, and the similar experiments of many others, send a deeply optimistic message: we don't need to fear the end of fossil-fueled industrial civilization. Instead, we can embrace this change, and help it come into the world.

CHAPTER 10

Slow Travel

The miracle is not to walk on water.
The miracle is to walk on the green Earth.

— Thich Nhat Hanh

It's possible to travel without relying on fossil fuels. I enjoy slow travel more than flying, even though it requires more planning and time. Slow, low-energy travel gives me time to think. It gives me a chance to connect with the land and the ocean—and I dearly love this planet—and with the people I meet along the way. It makes me more grateful for the time I have with the friends, family, or colleagues I'm visiting. And it's guaranteed to come with a healthy dose of adventure. It's not difficult, but it does require a shift in mindset.

Bicycle travel

In Chapter 8, we discussed the bicycle as a vehicle for local travel. Bikes are also practical for long-distance travel. A slow long-distance bike trip exists in an entirely different mental space than a fast airplane trip, or even car trip.

I once rode from Denver to Chicago over 16 days. I'd just finished a summer job building trail in the Sangre de Christo mountains. A few months earlier, I'd graduated from college, and I had no idea what to do with my life. So I decided to bike to my parents' house.

But first I needed a bike. I hitched a ride up to Colorado Springs. All I could find there were mountain bikes. So I hopped a

Greyhound up to Denver. After two days of hitching rides to bike stores—and two nights sleeping in a wooded margin behind one of the stores—I found the perfect bike, a heavy-duty touring bike on sale at REI for half price. It's the bike I ride to this day, an old friend, the Beast. I bought some rear panniers, a helmet, a rear-view mirror that clipped to my glasses, some spare tubes, and a small pump, and set off.[1]

To properly tell the story of this adventure would require a book of its own; here I can only give some flavor. The constant backdrop was pedaling on nearly abandoned two-lane highways from one small town to another. As I approached each town, I could see its water tower from miles out across the plain. One day early in the trip, I spent hours at a town's library, reading every book they had about bikes and bike maintenance. I only went 40 miles that day. Most other days, I biked around 100 miles.

Each day I'd pedal and pedal, and as the sun sank low, I'd start to look for a place to sleep. I'd head down the road in the failing light, figuring I'd go a ways and then settle in with the chirping crickets somewhere. Sometimes this is what I'd do. But more often than not, hospitality materialized from the ether, in one form or the other. I suppose you could consider a safe, cozy spot among the crickets a form of hospitality, anyway.

One night I knocked on a farmhouse door on the outskirts of a town, asking to sleep in a field, but was turned away (the only time this happened). I pedaled toward town. Before the road disappeared completely into the inky night, a red neon glow shone far ahead. This turned out to be the local watering hole, so I stopped and ordered a beer. The couple next to me at the bar was celebrating a birthday. They were curious about my trip, and we talked for a long while. They ended up offering me the couch in their single-wide for the night. The man was a cowboy—he rounded up cattle and drove cattle trucks.

His wife cooked breakfast the next morning as the cowboy barked orders imperiously. The atmosphere was toxic; in the light of morning, I realized both the man and the woman felt trapped.

I insisted on washing up. As soon as I finished, I thanked my hosts, stepped outside, and breathed in freedom as I saddled up and hit the road.

In cattle country, there were flies. Biting flies, and lots of them. Also long, rolling hills. To outrun the flies, I had to maintain a minimum speed up the hills. I'd sweat up those hills, and then roll down the other side ecstatically at speeds in excess of 50 mph, judging by the occasional cars which crept past and must have been going 60 or 70 mph themselves.

One morning, the Saturday of Labor Day weekend, I rode through a town hosting its county fair. I got talking to an older couple who invited me to spend the night at their place. They were empty nesters, and I reminded them of their son. The man was a grumpy farmer, and he thought I should be working at a job instead of tramping across the country on a bike. He taught me how to set the irrigation siphon tubes that sent the water from the Ogallala Aquifer onto his fields. I ended up staying two nights, to rest a bit and to avoid holiday drunk drivers.

That's how it went. Some nights I'd wash my "other" t-shirt and shorts in a public sink in a park, and sleep by the side of the road tucked somewhere hard to find. Other nights I knocked at the rectory in town, and I'd either be offered a spare room or a place to sleep on the grass; then I'd tag along for dinner at a congregant's house.

On my last night out, I stayed with Sharon's parents in western Illinois, a day's ride from my home town of LaGrange. At that time, Sharon was overseas, serving in the Peace Corps in Albania. I was hopelessly in love with her, but officially we were still just friends. It had been a long day. I'd taken a wrong turn in the late afternoon, which had tacked 12 miles onto an already long ride. Sharon's older brother gave me the third degree. I sat there, utterly starving but answering his questions with a smile. I think I passed, somehow.

As I pedaled, I earned each mile, and each mile had its own personality. Each day was an adventure. I was curious to see who I'd meet. I learned that I could trade the story of my trip, my optimism and openness, for a beer, a place to spend the night, a meal. I had no idea where I'd sleep at night, no deadlines at all. I was open to life, and life was open to me.

Driving on veggie oil

While I'm no fan of fossil fuels, I somehow happen to like old diesels. I like the way they clatter, and I like the way they can run on strange fuels, from kerosene to motor oil. Once, while returning from a backpacking trip in the Sangre de Cristo mountains with a friend in an old diesel VW Rabbit, we ran out of fuel. Half a gallon of kerosene, borrowed from a guy in a cabin, got us out of the woods.

Years later, shortly after moving to Altadena, I bought a 1984 Mercedes-Benz diesel for a song. She had 300,000 miles on her. I fixed a few things and converted her to run on waste vegetable oil (WVO) from restaurants.

In the future, as my awareness continues to deepen, I may find that I enjoy not owning a car more than owning one. I know people who choose not to own cars, and they get by just fine (though none have kids). But for now, driving on WVO is a good compromise, helping me live a new story within industrial society.

Making Maeby

Getting an old Benz to run on WVO isn't complicated. In fact, Rudolf Diesel designed his engine to run on mineral and vegetable oils in the first place. The crux of the problem is that WVO is thicker than diesel. Thicker liquids spray through nozzles in larger drops and in a narrower spray pattern. In an engine, this could cause incomplete combustion and engine damage over time.

From discussions on internet forums, I learned that the critical design question was whether to cold start on diesel and then switch over to WVO once the engine is hot (a two-tank system) or to cold start the car directly on the WVO (a single-tank system). The two-tank system uses engine heat to lower the WVO viscosity

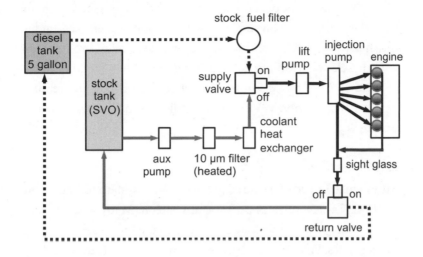

FIGURE 10.1. Diagram of Maeby's WVO conversion. WVO fuel lines are gray; diesel fuel lines are dashed; shared fuel lines are solid black.

to that of diesel fuel, while a proper single-tank system switches out the stock diesel injectors for injectors that work optimally with higher viscosity fuel.[2]

A single-tank system appealed at first. Superficially simpler, and with no need to warm up with diesel.[3] However, I could find no solid research on the proper injectors for running WVO in an old Benz. I began to realize that designing a proper single-tank system would likely require bench-testing various injectors myself, in order to replicate the proper spray pattern and drop size with cold veggie oil. Getting this wrong could ruin the engine, and then my low-energy experiment would end as just more stuff for the landfill. I decided on the two-tank system.

The other design decisions were relatively easy, and their implementation was straightforward (see Figure 10.1). I ordered two solenoid valves, a flat plate heat exchanger to transfer heat from coolant to WVO, and a large truck filter (heated by coolant in order to pass the oil more easily). Then I bought some rubber fuel hose, some hose clamps, and a spare set of steel fuel lines for a new five-gallon diesel tank which I situated in the trunk.

Those fuel lines posed the biggest challenge. Getting them from the junkyard involved lying on my back in broken glass under the shell of an old car on a blazing day in a field full of broken cars, with diesel fuel dribbling down around me—I felt like Mad Max.

And so Maeby[4] was born. I relished the process of creating her. And after 50,000 WVO miles, I still get a thrill when I switch over to veg.

Playing with fire

Meanwhile, I needed a source of grease and a system to filter it. At the time, I was a physics postdoc at Caltech. Many years earlier, an oceanology postdoc, long gone, had made biodiesel[5] from the dining hall's WVO. His abandoned apparatus—four 55-gallon steel drums, two with cone-bottoms, linked by various pumps and hoses—hulked over a corner of the machine shop in the basement of the chemical engineering building. I wasn't interested in cooking biodiesel, having opted instead for an upfront effort on the car to avoid the ongoing effort and dangerous chemicals required for biodiesel production. Preparing WVO fuel is far simpler: first you let it sit for a few days to allow tiny particles to settle, and then you run the settled oil through a ten-micron filter, discarding the bottom layer of sediment. With the blessings of the machine shop manager and the dining hall manager, I began collecting, settling, and filtering dining hall grease with the old equipment.

It's challenging to handle WVO without the occasional spill or leak. Once I made a mistake while hooking up a pipe, and I managed to pump a quart of oil onto the floor. I wiped it up with shop rags and thought nothing of it. A few days later, as luck would have it, I happened to be in the shop when the fire alarm went off. Smoke was coming out of the dirty rag can. My oily rags had spontaneously combusted! I hit the can with a fire extinguisher and then rushed it up the stairs and outside into the sun, into a small crowd of grumpy professors and blinking students who had evacuated their labs.

After this incident, I decided it wasn't a great idea to mess around with drums full of flammable oil under Caltech's chemical

Maeby in a world gone mad?

Maeby depends on a complex set of conditions to keep lumbering down the road. She needs parts, fluids, fuel, and the roads themselves. How hard would it be to keep her running in a "Mad Max" world?

- **Parts:** There are many other old Mercedes-Benz 300Ds around Los Angeles, so reusable parts could be had fairly easily, at least for the next decade or so. And assuming there were millions of empty, stranded cars, there'd be plenty of tires. But other parts—filters, seals, and the like—might soon become difficult or impossible to get, and would require creative improvisation.
- **Roads:** Our modern asphalt roads are oil-intensive. As the cost of oil rises, so does the cost of roads. As roads degraded in the Mad Max world, Maeby would become useless with her rear-wheel-drive and low clearance. But with almost no traffic, the roads might last for a long while. And Maeby would run just fine on desert hardpan.
- **Fuel:** Where does vegetable oil come from? The world's largest oil crops are oil palm, soybean rapeseed, and sunflower. An industrially farmed acre can produce an annual average of about 600 gallons, 50 gallons, 100 gallons, and 100 gallons of these oils, respectively.[6] Oil palms need a tropical forest climate, rapeseed likes cold temperate zones, and soybeans like moist grasslands.

In principle, I could cut down all of my beautiful fruit trees and grow sunflowers for oil. If I did this, I could probably get about three gallons in a year,[7] enough to drive Maeby 75 miles. I think I'll keep my fruit trees and hop on my bike instead—at least, until I run out of inner tubes.

engineering building. I now filter my veggie oil outside, in my backyard. This way there's less risk, and less mess.

The first epic trip: California to Illinois in the winter

Until a few years ago, we'd flown to Chicago to visit our parents every Christmas. My folks still live in the suburb just outside of Chicago where I grew up, and Sharon's folks live 90 miles to the west.

We'd never questioned this ritual. Could we trade in the airline for a 30-year-old experimental vehicle?

We'd already taken several shorter road trips in Maeby, to national parks in California and to a wedding in Arizona. All of these trips had been short enough for Maeby to carry all her fuel in her trunk. They'd also occurred in the heat of the Southwest summer, which is ideal for running on WVO. On a hot day, veggie oil flows easily, almost like diesel.

Los Angeles to Chicago and back in the middle of winter would be much more challenging; WVO congeals in the cold. Hoping to keep the WVO flowing through the fuel lines, I led a coolant loop under the car next to the WVO lines, and wrapped everything together with insulation. I led the loop up into the trunk and along the bottom of the WVO tank.

Another major challenge was even more primal: simply getting enough fuel. The total trip distance would be 4,000 miles, requiring 38 five-gallon cubies—about 1,200 pounds of oil! (Cubies are the plastic containers in which the new oil is sold.) I managed to lash six cubies to a roof rack. With another 6 cubies in the trunk and 18 gallons in the tank, we'd set off with 80 gallons of fuel, good for about 1,700 miles.[8]

Through a friend who worked for The Cheesecake Factory, I arranged to pick up WVO from franchises along the way. I also found a dude in Breckenridge, Colorado, who had some he could sell me.

With these preparations, we set off. The trip out was mildly challenging. We took the southern route, through Arizona and New Mexico, in temperatures slightly below freezing. While driving through Petrified Forest National Park, Maeby started to stutter. Fuel starvation. I'd been getting around 4,000 miles on filters, but this filter had only lasted 600 miles. I changed it, and we continued on. Outside of Oklahoma City, she started stuttering badly again, and I had to start running on diesel. This is a problem, given that the diesel tank is simply a five-gallon container in the trunk—without a fuel gauge. We'd been driving hard, and I was very tired, and we ran out of diesel late at night. I added the reserve diesel, but air was in the system. It took some doing to get her started, bleeding the fuel system in the cold night on the side of the freeway, exhausted and not thinking straight. This freaked out my family a bit, but I got her going again.

The biggest challenge for me at this point was not knowing what was keeping WVO from getting to the engine. Engine problems aren't always obvious when you're exhausted on the side of the freeway late at night. We limped into Oklahoma City and checked into a motel.

In the morning, I went to get WVO from the nearby Cheesecake Factory. The manager there informed me that the grease dumpster had been emptied the day before, but he'd had the collector save some for me. Of course, this was the bottom sludge—the stuff I discard in my own process. I was forced to put a tankful of diesel into Maeby in order to get to Illinois.

We had a nice two-week visit with our parents and siblings. Visiting over Christmas had typically been stressful for me, but it was noticeably less so after the slow travel, probably because I was more grateful to be there and less prone to take the visit for granted. A restaurant right under my dad's office had closed a few months earlier, and its grease dumpster turned out to be full of extremely well-settled oil. I gathered and filtered more than enough to make it halfway home to California.

On the way back, we took a more northern route through Denver, where we stopped to visit friends. We picked up enough fuel in Breckenridge, just over the pass above Denver, to make it home. But then the weather became extremely cold.

Driving through the high ski country, Maeby kept stuttering: her engine wasn't getting enough fuel. To get up hills, I sometimes had to switch her over to the tiny diesel tank. I finally realized that the main filter simply needed more heat to keep the fatty oil from solidifying.

Forced to run on the five-gallon diesel tank that lacked a fuel gauge, I did my best to keep track of mileage. But eventually I made a miscalculation and we ran out of fuel. I carried an emergency can of diesel, which I added, but the old gal wouldn't start. She had too much air in her fuel lines. I was under the hood for an hour, working the priming pump and having Sharon switch back and forth between WVO and diesel to maintain suction. My hand was bleeding from rubbing against all the extra fuel and coolant lines I've installed. The boys begin to cry. Finally, I felt the priming pump catch and move diesel fuel. I bled out the air, started her up, and we drove on.

That night found us driving on a cold and black stretch of Utah highway, slowly, with one headlight. Sharon and I were exhausted. We checked into a motel in Salina, Sharon in tears, talking about renting a car to get home. Everything seemed to be going against me.

This was the low point. Sometimes a moment comes, a moment in which I'm not sure I'm going to make it. It's an interesting moment, a teacher. I've learned to observe it objectively: what is this situation trying to teach me? The feeling of deep uncertainty, is that worth something? Reality is different than my expectations, and somehow this psychic friction causes suffering. But why should it?

There's no way to make the present reality other than it is; it can only be accepted or not accepted. Not-acceptance causes suffering. In that moment, *Sharon is disappointed in me, and this causes suffering. But why should it? I can't control another person's image of me, and why should her image affect my own happiness? But I'm also disappointed in myself. I've constructed a self-image in which I'm powerful, smart, and capable. But here I am, face-to-face with my limitations. I'm being taught that not accepting my own limitations causes me to suffer. I smile. With these observations in mind, I laugh at myself and at the situation. The deep darkness vanishes. It is what it is. Now what can I do to get my family safely home? Maybe this is a kind of practice for dying gracefully.*

I woke up before dawn to a −10°F morning. The second I stepped outside, I felt my boogers freeze: that kind of cold. I cranked Maeby until her battery died, undone by another preparation I hadn't made: an engine block heater. I swallowed my pride and called AAA. Soon the fossil fuel cavalry arrived in a gleaming diesel tow truck. We jumped Maeby, and after much cranking (and just a touch of ether), we got her running.

Later that day, on the climb outside of Barstow, California, the air conditioning belt let go while Sharon was driving; no big deal. Half an hour later, the power steering belt let go. I figured the extreme cold must have fatigued the old rubber, and I began to wonder about the alternator belts. Shortly after nightfall, we made it home, with no power steering and one headlight. I felt proud.

The trip was 4,400 miles, and we'd burned 35 gallons of diesel.

Long-distance WVO trips become a normal part of life

After learning from that experience, we got much better at taking long trips in Maeby. Our next trip was up to Portland, Oregon, then to Chicago, and then back home: an easy 6,030-mile trip in the *summer*. We burned 2.5 gallons of diesel, getting over 2,400 miles per gallon of diesel.

I'd installed a hitch, and Maeby pulled a trailer of fuel. We left Altadena with over 1,000 pounds of filtered WVO that I'd stocked up, 150 gallons in 30 plastic cubies. One of the amazing things about driving on WVO is how I become aware of the sheer mass of the fuel, since I have to lug it and pour it all myself. All that fuel disappears over the course of the journey as if by magic, right into the atmosphere. I sometimes wonder whether we'd have climate change if people had to lug their own fuel, or if CO_2 was hot pink and easy to see—either way everyone would be aware of just how much stuff they were dumping into the atmosphere. As it is, we pump it effortlessly, invisibly, and mindlessly into our tanks, and it comes out our exhaust just as effortlessly, invisibly, and mindlessly, allowing us to avoid knowing the material nature of our fuel, its sheer bulk. Out of sight, out of mind.[9]

As always, people along the way were fascinated by my journeys on veggie oil. I'd answer their questions, and most of them would tell me something to the effect that although they understood the severity of climate change, they didn't see a way to stop burning fossil fuels in their own lives.

During these trips, my awareness of our nation's addiction to fossil fuels deepened. On another summer trip (4,600 miles on less than a gallon of diesel, six nights camping, one night with friends, almost no money spent), we saw train after train of black tanker cars along the Nebraska highway. The few trains that weren't carrying Bakken crude were carrying coal. In the desert of Utah, near Grand Junction, Colorado, we saw fracking well after brand new fracking well along the interstate. One evening we followed a buried natural gas pipeline a mile into the backcountry, where we unrolled our sleeping bags and slept.

Once, while changing Maeby's WVO filter at a truck stop, I met a gasoline tanker driver fitting his hose to the port in the tarmac. I walked up and said, "I guess that's what keeps everything rolling."

The man, who wore a jumpsuit with *Dean* embroidered over his heart, looked at me kindly and said, "Do you know what would happen if we stopped the supply for a week?"

I said, "We'd all starve to death."

"That's right, we'd all starve to death. We're addicted to this stuff." Dean is all too aware of global warming, and feels helpless to stop it. He lives in Las Vegas and rides his bicycle the two miles to and from his job. The other drivers don't understand this strange behavior; Dean says they find it unnatural for someone to ride a bicycle two miles instead of jumping into a pickup truck. Dean had been driving and piping gasoline since 1976, but when I met

So you want to drive on grease?

In case you're interested, here are a few tips from my experience.

- Thirty-year-old cars are cheap to buy but costly to maintain. To avoid paying a fortune to mechanics, I work on Maeby myself. But this takes time away from other things.
- For a consistent source of grease, build a network. Partner with a nearby restaurant or two. You have to pick up reliably; join forces with other local greasers.
- Restaurants put their used oil back in the five-gallon plastic cubies for me, making it easy to collect. I can also collect from bins using a drum pump strapped to the side of a plywood-reinforced plastic bin and a length of vinyl tubing, but this is more work.
- Taking WVO out of a bin that's owned by a commodities company is theft, whether or not you have permission from the restaurant owner. Ten or twenty years ago, people often did this, but I wouldn't advise it.
- I settle the WVO for at least a week before pouring it through a one-micron sock filter and into clean five-gallon cubies.
- I add about a half teaspoon of algae growth inhibitor to each newly scavenged cubie of WVO.
- If I had it to do over, I'd keep the stock tank for diesel, and add a second Mercedes tank (from the junkyard) for WVO. I'd run fuel gauges for both tanks.
- Transporting WVO legally may require registering your vehicle for this purpose—and paying an annual fee.

What does it mean for a human
to "own" a mountain?
The mountain laughs at the notion.
Each chuckle takes it a million years.

him, he was looking forward to retirement in a year, to no longer living so deeply in the shadow of fossil fuel.

We camp along the way. In some parts of the country, it's challenging to find good family camping options. I dream of a world where every acre isn't spoken for, where traveling and camping along the way is an accepted, normal thing to do. Where do we get this strange notion that we can own land? Native Americans in the 19th century found it bizarre. What does it mean for a human to "own" a mountain? The mountain laughs at the notion. Each chuckle takes it a million years.

During another trip to Illinois, I worked as a server on a ten-day meditation course that my mom sat, her first. On the way back, we slept in Arches National Park, a mile off the road. Heading back to Maeby an hour before sunrise, we walked under Sand Dune Arch, quietly rising above us against the stars and the eastern glow. As I walked among the sandstone fins, with Sharon and the boys in the church of nature, I felt comfortable, welcome, and at home.

This is one of the great benefits of slow travel: a visceral sense of connection, that everywhere is home. Everywhere I go, I'm home—the whole Earth is my home.

Taking the train

I'm writing this section in the lounge car of the Southwest Chief, an Amtrak train that crosses the plains and mountains between Chicago and Los Angeles on a daily schedule. I'm on my way home from Christmas in Illinois. I love the train, and so does the rest of my family. As I write this, we're passing by the place where I fell in love with the wilderness as a youth, the place I built trail before riding my bike home to Chicago: the Sangre de Cristo Mountains near Raton, New Mexico. Yesterday evening, we passed within a mile or two of the Possibility Alliance, near La Plata, Missouri, an inspiring experiment in living without any fossil fuel whatsoever; I hope to visit the next time I pass through.

I spend most of my time in the lounge car, where I get a tremendous amount of work done; I go back to my seat to sleep at night. We actually sleep fairly well in the huge seats, which are more affordable than sleepers and result in less emissions. Even in coach, the four of us are emitting more CO_2 than if we were driving a 50 mpg car (see Chapter 9), and much more than if we were driving Maeby. But it's a nice change of pace. And as it's winter, I don't need to worry about Maeby's fuel congealing, or snowed-out mountain passes.

We prepared meals for the train. Sharon baked scones for breakfast and made a frittata for dinner, and I made *baba ganoush* for lunch. The trip takes about 40 hours, but Sharon and I get so much work done and enjoy ourselves so much that it feels, if anything, too short.

Crossing the ocean on a sailboat

Once I crossed the deep ocean on a small sailboat. This was a passage from Bermuda to New York City while I was a graduate student, and it took a week.

It wasn't technically travel, though. I flew one-way to Bermuda to help sail back a Jeanneau 36 that had just completed the Marion-Bermuda race. Since high school, I've had a dream to spend a few years sailing. On this trip, I tested the waters.

There were five of us on board. I stood four-hour watches with the three other hands. Crossing the Gulf Stream was eerie and beautiful. My first night on watch in those waters was quiet. With my watch mate asleep on the bench in the cockpit, I was the only person awake on an unknowably vast expanse of ocean. Bioluminescence scintillated in every splash and left a long glowing wake. The stars stretched from horizon to horizon. As the seconds turned into hours, the rocking boat lulled me into reverie.

> In this enchanted mood, thy spirit ebbs away to whence it came; becomes diffused through time and space...forming at last a part of every shore the round globe over.
>
> — HERMAN MELVILLE, *Moby Dick*

Suddenly the depth gauge, which had been reading infinity, started reading 30 feet, 20 feet, 25 feet...I rubbed my eyes. Despite my brain's knowledge that we were in the middle of the ocean with no land anywhere, I started to sweat. How sure was I that there was no land around? Could it be a submerged container? A whale? Dolphins? A few heartbeats later, the depth gauge started reading infinity again. I'll never know the cause. The ocean is like this.

Maybe the ocean is so vast and featureless that the mind starts to feel like anything can happen. It's no wonder to me that sailors tend to be superstitious.

The last two days of the trip gave us moderately heavy seas. It was quite a feeling to steer a sailboat on a beam reach with a huge wave towering up right behind me, far over my head. The boat would rise up and then surf thrillingly down the other side, and repeat this again and again. I could feel the forces on the vibrating rudder transmitted through the cables and into the wheel.

We made the New Jersey coast in the evening on the 4th of July, and we enjoyed the distant fireworks on the shore. The real show, though, was sailing up New York Harbor toward Battery Park with the city arrayed before us in all her predawn, fragile splendor. After crossing the ocean, the city was unreal, a mirage. It was as if I could see the buildings rising up and then falling down again, over the course of a few centuries.[10]

Hitching a ride on a container ship

In 2012, I rode a container ship to Hawaii and back. The SS *Horizon Spirit* sailed between the Port of Los Angeles and Honolulu every two weeks, bringing food to Hawaiians, right under the low marine clouds I'd begun working to decipher. For a year, the *Spirit* carried a suite of instruments for studying marine clouds. A newly minted atmospheric scientist, I was on board to learn about these instruments, to help monitor them, and to launch weather balloons from the bridge. The round trip took 12 days.

I parked Maeby at the Port of Los Angeles and passed through security. Enormous gantry cranes were loading the even more enormous *Spirit* as I walked up her gangplank. Everything was humming on fossil fuels, from the reefers, to the cranes, to the trucks, to the *Spirit* herself. Crew members of container ships fly to meet their ships for one-month stints. One of the *Spirit's* officers lived in India and flew home six times per year.

I slept through the predawn departure. When I woke, we were ghosting eerily through heavy fog. The creaky silence was shattered occasionally by the ship's massive air horn. I went to the bow of the ship and looked down. Dolphins were swimming with the ship and leaping into the air.

In the afternoon, I visited the engine room. It was hot and noisy, with two massive steam boilers powering two massive steam turbines which turn a massive propeller shaft. The propeller is called the "screw" because it screws through the water. The chief engineer told me it slips by only 5%.

To get to Hawaii and back, the ship burns about 7,000 42-gallon barrels of a low-grade fuel called "bunker C," which is the tar left over after an oil refinery has taken out everything else. A good run, which means departing LA on time in the early morning, burns about 6,600 barrels. If loading is slow and departure isn't until noon, the ship needs to sail faster to keep schedule, burning 7,300 barrels.[11] The sulfurous engine soot seeds clouds, creating ship tracks like those in Figure 10.2.

Does my voyage on the *Spirit* count as low-carbon travel? The RMS *Queen Mary* 2 emits 0.4 kg CO_2 per passenger mile,[12] comparable to a plane. But when hitching a ride on a container ship—so long as the ship's cargo isn't reduced to accommodate passengers—it's reasonable to account for the passenger's emissions as a mass fraction of the cargo. When looked at in this way, my emissions for the trip were 0.003 kg CO_2 per mile,[13] 1% of the emissions had I flown. That said, the biosphere would be far better off without container ships plying its seas. If some visionary entrepreneur brings back tall sailing ships, I'd sign on in a heartbeat.

In any case, I thoroughly enjoyed the trip. I finished an immense amount of work on my first paper in atmospheric science. And for me, time on the ocean is spiritual nourishment, at once refreshing and mystical. I'd often walk to the stern of the ship, under

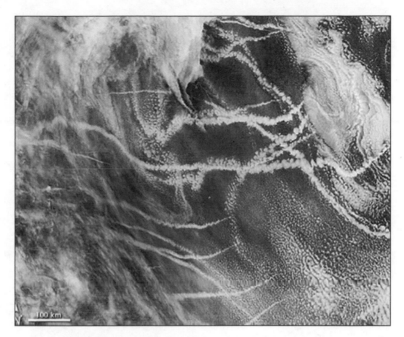

FIGURE 10.2. Ship tracks. Credit: NASA

⅕ of a mile of creaking and groaning containers, to where I stood high above the ship's churning wake, alone. I was surrounded by ocean. It was beautiful, stunningly so. With the moon rising low on the horizon, on that huge ship, the planet was very clearly a planet. And the moment was clearly a drop in a vast ocean of time.

A new story: Slow travel

The end of the era of cheap air travel may be approaching. As fossil fuels get more expensive to produce, air travel will get more expensive as well. If society begins to value the biosphere, it will implement a carbon fee, further driving up the price of plane tickets.

Regional conferences will become more important, as will remote conferencing technologies. Fewer of us will have the resources to fly across the globe for weddings, and we may move away from friends and family less frequently. Extended families may begin to regroup. It's wonderful to live close to loved ones, especially when they get old.

The year after that first epic trip Maeby, Sharon and I decided not to make our annual Christmas pilgrimage to Illinois. Although I missed everyone, it was also wonderful to shed the burden of holiday travel, the intense expectations that come with Christmas visits. I had many good phone and video conversations with my parents and sisters, including about someday moving closer to one another. The members of my extended family may even have taken each other less for granted.

Slow travel is powered by creativity instead of fossil fuels. There are so many other ways to go about it. You could build a WVO-powered motorcycle that gets 120 miles per gallon and drive it down to Tierra del Fuego, refueling with grease from empanada stands along the way. You could build a wood gasifier and drive on wood chips. Maybe you could even travel cross-country in a WVO-powered self-launching motor glider, harnessing the power of updrafts.

Or, you could fully experience the simple miracle of being on this Earth: you could walk.

CHAPTER 11

Meditation, a Foundation of Change

*We don't do sitting meditation
in order to become a Buddha.
We sit to be happy.*

— Thich Nhat Hanh

Whereas the immediate physical cause of global warming is our greenhouse gas emissions, the deepest underlying cause of the broader predicament is our *wanting*. Wanting drives both consumerism and overpopulation. We endlessly want more: more money, more sex, more prestige, a faster car, a bigger house, fancier furniture. And when we get something we want, the relief from our desire is fleeting. In a short time, we again want more. Wanting is a bottomless pit.

Not only is our wanting straining the biosphere to its breaking point, it's causing us to suffer. When we want something, it's because we're not satisfied in the present moment. Indeed, the entire purpose of the advertising industry is to cause us to feel unsatisfied with what we have. When we want, we feel agitated, unable to appreciate the miracles right in front of us. This dissatisfaction is *suffering*, but we may be so habituated to it that we don't recognize it. In this chapter, we'll discuss a simple practice that allows us to escape our wanting: meditation.

Meditation is a *practice*, not a religion. Remarkably, it requires no mystical revelation, no blind faith, no spiritual conversion. You become your own teacher. You learn by observing yourself in a straightforward way.

Many activists feel that sitting in stillness is a waste of time, but this isn't my experience at all. On the contrary, meditation makes my actions more effective by making them less full of ego. Meditation also bridges the disconnect I mentioned in Chapter 1, the gulf between what we know to be right and what we actually do. The intellect isn't capable of crossing this bridge. For these reasons, daily meditation is the foundation of my personal response to our predicament. It allows me to become happier, even as I carry a deepening awareness of the unnecessary suffering we're inflicting upon the biosphere, and ourselves.

The mind's basic habit

It's incredible that most of us go about our daily lives without ever stopping to observe our minds—how they work, and how suffering and happiness work. Most of us remain ignorant about our own minds, from birth to death.

Once you start observing it, you'll quickly notice that the mind is constantly reacting to pleasant and painful sensations in the body, even in sleep. Every fidget, for example, is a response to a sensation somewhere. *When there's a pleasant sensation, the mind wants it to continue. When there's an unpleasant sensation, the mind wants it to stop.* This is the mind's basic habit.

This subconscious habit, in turn, causes negativities like anxiety, jealousy, anger, hatred, and depression. For example, when we're anxious, it's because we're afraid of future unpleasant sensations, or of missing out on pleasant sensations. When we're angry, it's because we think someone or something is preventing pleasant sensations or causing unpleasant sensations.

It's impossible to deal with these negativities through willpower alone. I can't consciously will myself never to get angry, because anger is an instinctive, subconscious reaction that takes me by surprise and overpowers my conscious will. Instead, I need a practice capable of changing my mind's basic habit.

Our natural tendency is to try to run from suffering by constantly seeking pleasant sensations, without realizing that doing so only reinforces the mind's habit. We try drinking, drugs, sex, money, work, TV, getting on airplanes—all to no avail. Because every sensation, no matter how pleasant, sooner or later passes away. Our society is built around chasing happiness through consumption. But lasting happiness can never be found in this way.

We can't escape our minds. They follow us wherever we go. A more practical option is to make them good places to be.

Stillness and silence

Before we get into the nuts and bolts of meditation, let's take a moment to consider *stillness*. People in our society are always running, anxiously putting out fires in their lives. They keep running, mentally if not physically, right up to their last breath. But without stillness, it's impossible to know who we are and what we want out of life. If we don't know these things, we're just running pointlessly.

> Silence is not the absence of something, but the presence of everything.
>
> — GORDON HEMPTON

Since we're always running like this, when we *do* get an opportunity to sit in stillness, it's disconcerting. We want to get up and start running again. The stillness can be frightening to us, because when we're still, we may come face-to-face with our suffering. Being still, at first, is like riding a bucking bronco. But to come out of the suffering, we need to face it; and to face it, we need to be still. Stillness takes courage.

Our constant running is tied, bidirectionally, to our fossil fuel addiction: fossil fuel makes us run ever faster, and our running makes us love fossil fuel. We run noisily. We surround ourselves with television, even in public places. We put earbuds in our ears. The sound of cars and freeways is everywhere. In our houses, even when we think it's quiet, the refrigerator will switch on.

I long for quiet places, but they've become difficult to find. I'll put on a backpack and walk in the backcountry for days, only to hear a nearly continuous roar of airplanes overhead. I seek out

The goal is to change the mind's deepest habit,
after all, and the mind is a slippery fish.

quiet, because to me it's as beautiful as the night sky in the darkest mountain wilderness. This beauty is part of the vision I hold for a world without fossil fuels.

The practice

There are many ways to meditate. Here, I'll explain the simple and concrete technique that I practice, but this is *not* a meditation lesson. It's merely a description. I know that I couldn't have established my practice at home from a written description.

If you want to learn, it's best to go to a meditation center.[1] Although the practice is simple, the mind is a slippery fish. It will strongly resist your attempts to change its deepest habit. The mind will think of surprisingly creative and persuasive excuses to stop sitting and get up. To learn meditation, you need a firm commitment, a quiet place free from responsibilities, a sustained, single-minded persistence over at least ten days, and a teacher who can answer any questions that come up.

> If your practice does not bring you joy, you are not practicing correctly.
>
> THICH NHAT HANH

How Sharon and I became meditators

My upbringing, as a not-so-devout Roman Catholic in the suburbs of Chicago, was about as far from the mystical East as it's possible to get. But in college, I discovered Indian music and seriously considered becoming an ethnomusicologist. After college, while teaching high school physics and astronomy in rural southeast Massachusetts, a musical friend from India called me to say he was attending something called a "vipassana course" a few hours away. The logistics for meeting up didn't work out, though, and I didn't follow up on the vipassana thing.

A few years later, while having brunch at the Cornelia Street Café in Manhattan, another friend casually mentioned that she was going to serve at a meditation course for ten days. She'd be cooking food for the meditators and doing their dishes. Intrigued this time, I found an essay entitled "The Art of Living"[2] which resonated strongly with me; the basic idea was that we're suffering, but through direct observation (which I'd never thought to do), we can figure out the cause of our suffering and how to eradicate it.

Sharon was also at that fateful brunch. During her time in the Peace Corps (while I was teaching high school), she'd read Robert Thurman's *Essential Tibetan Buddhism*[3] along with other Buddhist texts, and tried to teach herself meditation. However, she found learning meditation from books to be impossible. She was also eager to try vipassana.

So we signed up for the next available course, headed to the Port Authority bus terminal, and hopped a Greyhound to Shelburne Falls, Massachusetts. Over the ten days, we'd live segregated by sex; we agreed that if either of us thought we'd stumbled into a cult, we'd sneak across to the other side and we'd leave together. But as the course progressed, I grew more and more convinced that this was the real deal. During the course, I experienced the highest highs and the lowest lows—and had my first taste of equanimity. I was forever changed, and so was Sharon.

Experiencing change

You can always follow your breath. Observe the flow of air through your nostrils with each inhale and exhale. Notice the slight tickle on your upper lip, the warm or cool flow of the air. Witness these subtle sensations as you breathe naturally. As long as you're living, your breath will be there for you—in any situation. Working to follow the breath, for longer and longer periods, sharpens the mind and prepares it for the core practice of meditation.

This core practice is an extension of the breath awareness: to observe the normal, everyday sensations throughout the body (pain, moisture, heat, tingling, etc.) without reacting to them. The goal is to observe and *do nothing—to break the habit of reaction.*

In life unwanted things happen and wanted things don't happen. We can't control this. But we can learn to dampen our reaction. Toddlers, and even most adults, immediately react with a tantrum. They suffer immensely and cause everyone else to suffer as well. A meditator, though, will smile calmly—because *this too shall pass*—and then, with a clear mind, get on with the work of making a better world.

Being quietly aware of the sensations on my body, I see that they come and go on their own. They're constantly changing. Experiencing this impermanence, I develop equanimity. What's the point of being attached to this or that sensation, when sensations are impermanent? Why multiply my suffering through such meaningless attachment? Direct, sustained experience of impermanence gradually breaks the habit.

As best I can, I remain alertly aware as I patiently scan up and down, from the top of my head to the tips of my toes and back, examining every area of my body as I pass it. Suppose that in this moment, my attention is on my right shoulder. I quietly observe whatever sensation my right shoulder is experiencing. It doesn't matter whether it's pleasant, painful, or neutral. It always arises, and then passes away—impermanent. Then I move to my right upper arm, and repeat. Then to my right elbow, and so on, covering the entire body systematically. If my mind wanders away from this exercise, which it often does, I bring it back without disappointment or frustration. This, too, is part of the practice.

Every sensation is a tool for coming out of the habit of blind reaction. Observing pleasant sensations without reacting dissolves negativities associated with greed. Observing painful sensations without reacting dissolves negativities associated with hatred. And observing neutral sensations without reacting dissolves negativities associated with boredom.

By directly observing bodily sensations, I experience that they are impersonal, and that attachment to them (projecting "me" into them) causes suffering. As I sit still, sensations come and go on their own. By experiencing their changing nature, I experience the changing nature of this mind-body composite I call "me." Experiencing it, I realize that I'm impermanent. The notion of "I" gradually dissolves.

The ego is like grime on a window. Once the window is washed clean, the underlying reality of connection shines through like the sun.

Other aspects of the practice

This practice—the equanimous observation of change, in the framework of bodily sensations—is called *vipassana*, which means "seeing things as they really are." It's the practice Gotama (the Buddha) discovered and used to come out of his own suffering. There's nothing religious about the practice. Buddha wasn't interested in creating sects labeled "Buddhist." He simply taught people a practice he'd discovered that allowed him to come out of suffering.

Buddha taught two other helpful practices. The first is breath-awareness, *anapana* (which means inhalation-exhalation). The second is wishing all beings to be happy, a meditation of sharing peace, *metta bhavana* (which means cultivation of loving-kindness).

About once a year, I go on a ten-day meditation course. The first three days are devoted to *anapana*. The mind is naturally wild and unstable, wandering randomly from thought to thought, wallowing pointlessly in regrets of the past or anxieties over the future. Practicing breath-awareness gradually tames and sharpens this wild mind, preparing it to practice vipassana more effectively.

At home I meditate for an hour after waking up and an hour before going to sleep. After these sits, I practice *metta* meditation for a few minutes. I remain aware of sensations while wishing other beings to be happy. If there's anyone for whom I've felt ill will, I might direct my *metta* to that person. I put myself in that person's situation, understand his or her hopes and suffering. As I'm practicing this, I'm aware that others are also practicing it. For me, the realization that there are other beings selflessly wishing for everyone to be happy and peaceful is a great source of strength. *There is help, and love, and goodwill out there.*

To support the overall practice, it's important to avoid killing, stealing, lying, sexual misconduct, and intoxicants. These things both arise from and encourage reacting blindly to craving or aversion, reinforcing the mind's basic habit and causing harm to oneself and to others.[4] They also agitate the mind, as I know well from my own experience. As I continue to practice vipassana, avoiding these harmful actions becomes easier. This isn't puritanism: there's no repressed desire. Instead, I simply experience that when I break these precepts, I suffer. So I prefer not to break them. The desire evaporates.[5]

Mindfulness is the capacity to remain aware of what's actually happening, within and without. As I continue practicing, the distinction between sitting meditation and other times is becoming less sharp. I try to maintain awareness of some bodily sensation or breath at all times I'm awake. Although I don't succeed at this yet, it helps me to remain mindfully in the present moment, whether or not I'm sitting in meditation.

Some benefits of vipassana

Although meditation teaches us to come out of craving, it's important to experience tangible benefits. Otherwise there's no way you could sustain sitting for two hours every day! I personally find that spending two hours per day is absolutely worthwhile. Indeed, if you're getting no benefit from your practice, it's a sign that you might be practicing incorrectly; in this case it can help to talk to a teacher.

What are some benefits? First, the stress and anxiety in my life evaporates. I'm less concerned with outcomes. If I let my practice slip, shortening and skipping some of my daily sits, anxiety creeps back into my life. When I start practicing steadily again, the anxiety goes away. My work is just as challenging while I'm meditating, but I'm able to see it for what it is: *not me*. With a detached mind, I work more effectively and make better decisions.

Second, I become unconcerned about what others think of me. As a nerd growing up, I desperately wanted to be seen as cool. Even as an adult, I craved that people would think about me in a positive way. This was mental slavery: when I felt someone thought well of me, I'd feel happy; when I felt someone thought ill of me, I'd feel unhappy. Now I simply do what I need to do.

Third, I suffer less from all types of negativity. I have less anger, less depressed feelings, less loneliness, less hatred, less craving for extramarital sex, less jealousy, less negative self-talk, and less impatience. This means less suffering for me. It also allows me to have more harmonious and satisfying relationships with my family and friends. Sharon and I fight far less, and far less intensely, than we used to. I hate no one, and I see how all my past hatreds originated in myself. Everyone is my friend. Even the thought that someone could be my enemy seems odd to me now.

Fourth, I feel more gratitude and connection. I see how precious life is. I feel more gratitude for the people in my life. I fight with them less, and it's easier to give them *metta*. More compassion and openness allows me to connect easily with strangers. People come up and talk to me. I smile, they smile.

You may or may not experience these benefits, and you may experience others. Take care not to *crave* some specific benefit, however, as this will reinforce your mind's habit of wanting. Just let your practice unfold over the years. You can't predict how this will happen.

Keep in mind, though, that there's only one measure of progress in this practice: *equanimity*. If you practice correctly, you will become more equanimous. This will necessarily make your relationships with others more harmonious. If you don't feel more equanimity and harmony with those close to you, you need to make some correction.

Meditation and the brain

Meditation feels like a fundamental rewiring of the brain. From a neurological perspective, this turns out to be the case. *Neuroplasticity* is the brain's ability to learn by changing its structure, rewiring itself as we learn a new skill, such as a sport, a musical

TABLE 11.1. A few of the physical changes in the brain and corresponding benefits due to variants of mindfulness meditation.

Physical change	Effect(s)
Larger hippocampus	Improved emotional regulation
	Reduced risk of depression
Larger temporal-parietal junction	Increased compassion
Smaller amygdala	Reduced anxiety
	Decrease in emotional reactivity
Thicker prefrontal cortex	Improved concentration
	Improved decision-making
Delayed cortical aging	Retention of memory and thinking
Rewired default mode network	Reduction in mind wandering
	Increased happiness

Credit: Peter Kalmus. For sources, refer to the text and endnotes.

instrument, a language—or meditation. I think it's interesting to take a detour and see what science has to say about meditation. My goals in so doing are to present meditation from a complementary perspective, and also to allay any fears you may have that meditation is some mystical or religious thing. It's just a practice. An incredibly valuable practice, but still just a practice.

Below I provide a brief summary of a few (though by no means all) ways the brain changes due to meditation (summarized in Table 11.1). I've chosen studies focusing on mindfulness meditation, a generic term signifying meditation that is similar to vipassana (i.e., equanimous awareness of bodily sensations).[6]

Significant regional changes in gray matter volume and density occur after as little as eight weeks of 30-minute per day meditation practice.[7] Meditators show significant increases in the left hippocampus, a region of the brain associated with emotional regulation, learning, and memory.[8] Conversely, people with major depression have a smaller hippocampus.[9] These two facts, taken together, suggest that meditation prevents depression—a hypothesis which other studies[10] (and indeed my own experiences[11]) confirm.

The right temporal-parietal junction (a brain region above the right ear), which is associated with empathy, compassion, and self-awareness, grows larger.[12] The amygdala, associated with anxiety and the flight-or-fight response and known as the "fear center," shows a *reduction* in gray matter due to meditation, and decreases in gray matter there have been correlated with decreases in stress.[13] Cortisol, the stress hormone, is decreased after a period of meditation.[14] Finally, the prefrontal cortex, associated with higher order brain functions such as awareness, concentration, and decision-making, becomes thicker with meditation.[15] The cortex usually shrinks as we age, which is why as we get older we experience memory loss and difficulty in thinking; but meditators didn't show the usual decrease in cortical thickness with age, implying that meditation prevents the brain's usual atrophying with age.[16] And there's evidence that meditation prevents age-related cognitive decline.[17]

Meditation also affects the brain's "default mode network," the neurological wiring responsible for the default behavior of our minds, which is to wander.[18] Other studies have shown that a wandering mind is correlated with unhappiness.[19] The main nodes of the default mode network are less active in meditators, both while meditating and while not meditating (in other words, the changes persist in daily life). A new network becomes active instead, coupling parts of the brain associated with self-monitoring and cognitive control (the posterior cingulate, dorsal anterior cingulate, and dorsolateral prefrontal cortices).[20]

Researchers have also demonstrated that meditation reduces pain,[21] cures insomnia,[22] boosts the immune system,[23] and lowers blood pressure.[24]

A universal path that isn't for everyone

While anyone can practice vipassana, not everyone will feel drawn to it. Many people simply aren't seeking to change themselves, to come out of their suffering and live a more harmonious life. Other people are seekers who are drawn to different paths.

Other practices that lead people to actually increase their equanimity, compassion, and happiness are also good practices. If the practice increases your ability to respond to the negativity of others with compassionate love, and your relationships with others grow more harmonious, then the practice is helpful.

Meditation and our predicament

Vipassana meditation is a fundamental part of my response to living in a warming and overpopulated world undergoing rapid transformation. As part of my daily work, I look directly at the truth of global warming, and what it's doing to the inhabitants of the Earth. Meditation gives me the strength and the courage to keep interacting with this truth, as it is—not only to cope, but to be happy and as effective as possible in enacting positive change.

Meditation drives my ability to change myself. Because of meditation, my desire to consume has greatly diminished. I no longer have any desire whatsoever for vacation homes, sports cars, or private jets. I'm happy with enough.

By experiencing connection, the last thing I want is to intentionally harm other beings. By experiencing how all actions have effects, I wish to perform only actions that are good for me, good for others, and good for the biosphere. Meditation has been, and continues to be, the key to aligning my actions with my principles.

The ability to see reality as it is (and not how I want it to be) leads to acceptance of my situation, which leads to appropriate action. When we're in danger, we need to recognize the danger correctly before we can act in our best interest. If we delude ourselves, denying the existence of the danger, having false hopes, or perceiving the situation through the lens of a preexisting mindset, we'll act accordingly.

Meditation develops gratitude, something that's critically lacking in industrial society. When I view food, water, energy, community, and time as precious things I'm grateful for, I actively seek ways to avoid waste. This becomes a joyful task.

Like you, I'm a fold in the universe that took
this human shape for some time and will change,
like foam on the surface of the ocean.

Meditation also develops compassion. This is particularly relevant to our predicament: if we feel hatred toward the oil company executives and the ultra-rich, for example, we are still planting seeds of hatred. If we manage to change the social regime, but we do so through violence, greed, and ignorance, only more violence, greed, and ignorance will grow out of our actions. History will continue to repeat.

Finally, meditation reduces conflict and makes living in community much easier. As global warming intensifies, and global systems break down, we'll increasingly turn to our communities. By allowing us to observe and temper our automatic responses, meditation facilitates discussion across ideological divides, which is more important than ever. The ability to interact with others, free from the destructive narcissism of the human ego, is as helpful to the members of a family as it is to heads of state.

The experience of real happiness and love is there for each of us all the time, underneath, but it gets clobbered by the intensity of the noise of "me," the noise of fear, anger, selfishness, jealousy, hatred, regret, hopelessness, and *wanting*. It's impossible to truly love with all that noise. We'll come out of the nightmare of history only after enough of us learn how to generate love for all beings.

The sages have been *saying* "love one another" through the ages. But without a practice, a simple technique that works, these too easily become empty words. In vipassana, we have just such a simple, concrete practice, available to all, that allows us to actually come out of selfishness and hatred, that allows us to come out of this illusion of "me." What a wonderful thing!

As I directly experience the constantly changing nature of my body and mind, my ego gradually dissolves. As my ego dissolves, my experience of connection deepens. I *experience* that there's no "me," only a flow. I *experience* that I'm a form that arose out of these quarks and leptons, these subatomic particles which physicists believe are themselves ultimately just vibrations. This form will come apart again. Like you, I'm a fold in the universe that took this human shape for some time and will change, like foam on the surface of the ocean. What once seemed solid and permanent was actually a flow dependent on the food we eat, the existence of our parents, the existence of the biosphere, the existence of stars.

And then, when the illusion of separation dissolves, what's left is love. A love without attachment, arising from a tiny conscious piece of the universe. A selfless love that positively radiates kindness, compassion, joy, and equanimity.

CHAPTER 12

Reconnecting
with Mother Earth

*To forget how to dig the Earth
and to tend the soil is to forget ourselves.*

— MAHATMA GANDHI

Most people in industrial society neither know, nor care, where our basic human needs come from—our food, shelter, and water—or where our waste goes. The infrastructure and belief systems in our society create the illusion that we're separate from nature, even though we're connected down to our atoms. We are embodied beings, mammals in a biosphere. We breathe air, drink water, eat food.

Growing up in the Chicago suburbs, I had no idea where food came from, and I never thought about it, not even while eating. But I now see that food connection is essential for reimagining humanity's place in the web of life.

In 1900, 41% of people in the US were growing food; by 2000, the number was less than 2%.[1] May this chapter inspire you to get your hands dirty.

Growing food (becoming)

Growing food is at once a revolutionary act and a skill that may come in handy if industrial civilization declines. It's also one of the most satisfying changes I've made to my life. I was hooked after my first homegrown tomato.

I didn't get to know plants until I was in my mid-thirties. My parents weren't into gardening. When I left for college, I began a long urban journey living in tiny apartments in some of the world's largest cities. I felt disconnected, cut off from the Earth.

When we moved to California, I felt an immediate connection to the large and generous avocado tree that shades our house. The tree drew me in, gently. At first, I continued paying the mow-and-blow crew the previous owner had used. Every two weeks, they'd come with their noisy machines and fill up the green waste bin. But I prefer quiet to the sound of leaf blowers, so I stopped the yardeners. I put up a hammock under the avocado tree and gazed into its leaves.

Three of my neighbors were experienced gardeners. Jimmy and April, my neighbors across the street, were generous with their knowledge as I began to learn about plants. They were the first on our block to rip out their lawn and plant food in their front yard. Their garden still inspires me; I often take a moment to see how it's doing before biking to work.

Ruben, my next door neighbor (and a dead ringer for Carlos Santana), had a green thumb and kept chickens. When he moved away, he gifted me two chickens and some potted plants. I became a chicken keeper overnight, and I had to figure out how to make the plants happy.

Now, many years later, I see that plants are beings. When I pick a leaf or a fruit from a plant, or even when I pull out a weed, I do so with awareness, respect, and gratitude. Every seed is precious, a tiny genetic blueprint of unimaginable complexity, evolved over billions of years, for a plant being that creates a miraculous gift: nourishing food that we can eat.

A few of my favorite plants

Gardening demands that you respect your land and your climate. What works for me in Altadena may not work for you. Talk to other gardeners, find out what works in your town, and plant what you like to eat.

With that caveat in mind, allow me to introduce you to a few of my favorite plants.

First and foremost, the fruit trees. Coming to California from Manhattan, I was amazed to find fruit trees all over the place. When I moved in, there were nine fruit trees in my yard, but I promptly killed the plum and the peach. This motivated me to learn the basics of tree care, and to plant bare root trees ordered through the mail. I now have two dozen happy fruit trees on my tiny, ¹⁄₂₀ acre plot. My favorite is the avocado, but I also love the orange, fig, pomegranate, and nectarine, which makes gorgeous flowers. These trees produce the best fruit of their kind I've ever tasted. The lemon tree keeps its fruit for half a year or so, during which time we cook with lemons daily.

I love having great homegrown fruit in season, and then taking a break until the next season. When fruit comes back the next year, the satisfaction is intense. This seasonal cycle is part of the pleasure of homegrown food.

Since I'm a lazy gardener, I also love perennial vegetables. We have a several beautiful artichoke plants in the front yard. During a good year, each plant can grow to nearly the size of a VW Beetle, and produce 20 chokes. (They're a delicacy simmered in olive oil, white wine, lemon juice, and garlic.) Arugula takes over my yard in spring and early summer, and I think of it as a perennial because it reseeds itself. I love its fresh spiciness, especially on sandwiches.

All homegrown food tastes better than what you can buy, and this is perhaps more true of potatoes than anything else I grow. A baked homegrown potato is food for the gods. My potatoes and tomatoes are both relatively vulnerable to pests, so healthy, complete soil is essential for strong plant immune systems. Crop rotation helps prevent problems from overwintering diseases. I also love growing chard, garlic, mustard, kale, cabbage, eggplant, sweet potatoes, peppers, rosemary, asparagus, blackberries, squashes, rhubarb. And sorghum is a grain crop that basically grows itself here, and is delicious cooked like rice.[2]

Over the years, I've had successes and failures; I don't stress over the failures. And I continue gleaning wisdom from my fellow gardeners at the community garden.

If 100% of the people were farming it would be ideal.

— MASANOBU FUKUOKA

The vibe: Easy does it

Gardening can stress you out if you let it. In terms of gardening skills, perhaps the most important is self-awareness. How do we relate to our garden? Is it a source of renewal or stress?

Labor-saving gardening practices

My first years of gardening involved much avoidable frustration. Here are a few lessons I've learned.

- **Irrigation:** Drip irrigation is essential in hot, dry climates like Southern California. Irregular watering stresses plants and reduces yields. Vacations reliably killed my garden, no matter the good intentions of the garden sitter. Installing a drip system turned out to be easier than I expected; like most things in gardening, drip irrigation systems can start small and grow as needed.
- **Perennials:** My perennials keep growing, year after year, with little work on my part. In addition to trees and perennial vegetables,[3] I like annuals that replant themselves. After moving to a new home, start perennials first as they take a few years to become established.
- **Mulch:** A thick layer of wood chips, shredded leaves, or straw, mulch sits on top of the soil like a shield. Mulch can cut water use nearly in half, and it reduces weeds and enriches soil. I like a four-inch layer around perennials. When I need mulch, I'll ask arborists working in the neighborhood for their chips. This saves them an

When I first started, gardening felt like too much work. It seemed that critters and pests were out to get me, and that nothing would grow for me. Eventually I developed soil appreciation. I've also become less attached to getting a crop. Plants come, and plants go. Sometimes they yield a good crop, sometimes they don't.

I've also developed patience for the hungry critters. This is their land as much as mine. They don't mean anything personal by eating "my" avocados or digging in "my" garden beds. Like me, they're just trying to make a living. Squirrels enjoy taking one bite

expensive trip to the dump; it also cuts down on landfill emissions (a win-win-win, a three-way gift).

- **Sheet mulch:** An easy way to prepare a grassy or weedy plot for gardening: loosen the soil with a garden fork, lay cardboard over the grass or weeds, and lay mulch on top of the cardboard. Keep moist.

- **Locale specificity:** It takes years before fruit trees start producing, so only plant what works where you live. Talk to your gardening friends. (For example, see the Great Altadena Fruit Project in Chapter 15).

- **Spacing:** Trees and other plants need adequate space. A rookie mistake is planting too close.

- **Tilling and double digging:** Some swear by this, but I don't bother. Too much work. And probably hurts soil communities.

- **Saving seeds:** I can save exact varieties, save money, and save in large quantities for giving away. Saving seeds is more important than you think—it's a key to building resilient communities around the world. Consider starting a seed library.

Fukuoka saw no such battle;
he saw a dance.

out of every avocado, but I don't mind; I just cut the bite-scarred area away as I peel the ripe fruits.

I've noticed that if I feel frustrated by critters, I suffer and they seem to do more damage. But when I feel compassionate toward them, I don't suffer, and they seem to do less damage.

It's important to recognize and avoid unnecessary work, to work with nature instead of against it. This saves time, effort, and lower back pain. The grandpa of permaculture, Masanobu Fukuoka, spent his life perfecting the natural farming of rice, citrus, vegetables, and winter grains. In his 1978 book *The One-Straw Revolution*, he described his principles of natural farming: no plowing, no chemicals.[4] Fukuoka relied on timing (seeding the winter crop before harvesting the summer rice), green manure (clover), straw (he returned everything except the harvest to the field), and a sprinkling of poultry manure (to help decompose the straw and make up for the harvested grain).

Fukuoka let the web of life do the heavy lifting, and his yields were routinely as high as those of any industrial farm in Japan, year after year. So why doesn't everyone farm like Fukuoka? First, natural farming requires a long-term outlook, as it can take a few years for natural systems to rebuild and stabilize. It's difficult to make the switch on a for-profit farm.

It also takes a deep, experiential understanding of the soil and the ecosystem—especially the weeds, insects, and insect predators. Fukuoka learned what worked on his unique patch of land by trial and error over many years. (He wrote, "I probably know more about what can go wrong growing agricultural crops than anyone else in Japan."[5]) In industrial farming as well as in many backyard gardens, we replace deep understanding with fossil-fueled technology. This seems more convenient: in this modern age, who has time to observe the web of life? But in not observing, we can destroy this delicate web, which locks farmers into an uphill battle using arsenals of heavy machinery, chemicals, and GMO seeds against natural forces.

Growing food should be built into
the elementary school curriculum
on an equal footing to
reading, writing, and arithmetic.

Fukuoka saw no such battle; he saw a dance. But his dance required patience to learn.

Finally, there's no need to put in a whole garden immediately. Mine has grown gradually over the years. When I feel like working in the garden, I do; when I don't, I don't. I'll take care of ugly or weedy patches when I feel like it. In spring I relish weeding a day or so after rainstorms, when the ground is soft and the weeds are young. When done in pleasant weather and with full attention, I know of few things as therapeutic and relaxing as weeding.

Could I feed my family by growing my own food?

It took me about five years to gain basic competence as an organic gardener. If necessary, would I be able to use the skills I've gained to feed my family?

If I had about an acre here in Altadena, with its year-round growing season, and made growing and preserving food my highest priority, I *probably* could. This assumes access to a few things: a reliable water source; established fruit trees and other perennial food plants; and a diverse community of dedicated gardeners with whom I could trade knowledge, excess produce, and seeds.

My experience matches that of Fukuoka, who wrote, "If each person were given one quarter-acre, that is 1¼ acres to a family of five, that would be more than enough land to support the family for the whole year. If natural farming were practiced, a farmer would also have plenty of time for leisure and social activities within the village community. I think this is the most direct path toward making this country a happy, pleasant land."[6]

It takes years to gain the skills and build the soil one needs to grow food, and to do so without strong community support would be impossible. Beginning to grow food because you're facing starvation would work about as well as beginning to sew a parachute because you've fallen out of a plane.

I used to have unthinking faith in the global industrial food system. Now I believe we have a responsibility to teach our children how to grow food, both to help them live satisfying lives, and to boost community resilience. Gardening should be right up there with reading, writing, and arithmetic in our schools: a basic literacy of food, plants, and soil.

Sustaining soil with compost (returning)

I'm learning to think of plants and soil as one system. When soil is good, plants grow well. When soil isn't good, pests and diseases move in easily.[7]

When I started gardening, the soil in my yard was impoverished. But buying plastic bags full of amendments from the garden store and schlepping them to my house seemed crazy. When I first heard about *humanure* composting, it also seemed crazy—but also too sensible not to try. Composting my own waste reduces CO_2 emissions, water usage, and water pollution. It's an elegant step toward closing the loop on food and waste. It's pure ecological alchemy, transmuting a major problem (polluted and biohazardous water) into a major blessing (fertile soil), naturally and without money.

> Hmmm. Where *can* a large animal like a human being find manure?
>
> — JOSEPH JENKINS, *The Humanure Handbook*

Plants give to us freely: cool shade, beautiful forests, building material, fuel, clean air, clean water, food to eat. It's no exaggeration to say we owe plants our lives. In return, we can give them a gift that's a pleasure for us to give: our poop and our pee. The biosphere is a beautiful system, indeed.

Composting my own waste is a meditative practice. It's the opposite of *out of sight, out of mind*. It reminds me that I'm an animal. It also feels subversive, which makes it fun.[8] I've done it for years, and it has become a natural, easy part of my life.

At first, humanure felt like another project, an addition to my life. Now I understand that it's a simplification. I've subtracted the flush toilet from my life. I've subtracted the need to buy bags of soil amendment. Once a week, typically on Sunday morning, I empty a bucket onto the compost pile. As I do so, I'm reminded of the alchemy taking place.

How to compost humanure

In composting humanure, microbes do all the real work. Humanure requires no fossil fuels and less than 5% of the water used by flush toilets.[9] I follow the system Joseph Jenkins describes in his *Humanure Handbook*.[10] The toilet is a bucket under a standard toilet seat mounted on a plywood box.[11] You can build one in a matter of minutes.

"Flushing" is accomplished with a few handfuls of leaves (shredded is best), horse manure (dried is best), or any relatively fine mulch (generally referred to as brown material). Jenkins uses sawdust from local sawmills, which is easy for him to source in western Pennsylvania. So long as you use adequate brown material to cover newly deposited poop, there's no odor.

In addition to the bucket, the system requires two compost piles, one for buildup and one for mellowing, which I wall in with scavenged wood pallets. I fill the buildup pile with buckets of fresh material. To do this, I move aside some topping brown material with a dedicated pitchfork, dump the bucket, and then replace the topping material. Then I rinse out the bucket with a hose and dump the water on the pile. If there's odor, I add more brown material.

After a year, I use the dedicated pitchfork to move the buildup pile into a second pile, the mellowing pile. There the compost mellows for a second year.

After two years, humanure has transformed into nutrient-rich compost. I add it to the garden and the orchard, laying it down on top of garden beds, under the drip lines of trees, and to any plant that seems to want some extra nourishment. The received wisdom is to restrict the use of humanure to fruit trees, and to never use it near vegetables. While this rule is certainly necessary for raw sewage, it's overkill for properly composted humanure.

Hygiene

To be safe, humanure must be properly composted. In practice, this means it must reach a temperature hot enough to kill pathogens.

It's worth noting that cholera pandemics were a fact of life, even in the US and Europe, until quite recently. A disease that spreads when sewage contaminates the drinking water supply, cholera claimed tens of millions of lives in the 19th and 20th centuries. Even today, more than 100,000 people per year die from cholera; and global warming is expected to give the cholera bacteria a boost.[12] Cholera isn't the only disease you can catch from sewage: feces from infected individuals can transmit other bacteria, parasitic worms and protozoa, and viruses.

However, proper humanure composting eliminates these dangers. Fecal pathogens are killed by heat over time, and humanure piles heat up due to heat-producing thermophilic microbes—good bacteria. The hotter the pile becomes, the less time it takes for pathogens to die: "complete pathogen destruction should be guaranteed" if all parts of a compost pile are maintained at temperatures of 144°F (62°C) for one hour, 122°F (50°C) for one day, 115°F (46°C) for one week, or 109°F (43°C) for one month.[13] This is the golden rule of humanure.

During the buildup phase, my pile holds steady at 110°F to 120°F no matter where I stick the compost thermometer. More importantly, the temperature in newly added five-gallon deposits climbs to 135°F to 140°F after two days, and stays up there a week or more.[14] After three days at this temperature, essentially all pathogens in my newly added material will be dead. All additional time is insurance.[15]

Turning the material by pitching it from the buildup pile to the mellowing pile provides a second round of purifying heat. I once noted that two days after doing this turning, the temperature in the newly mellowing pile climbed to 155°F. It stayed above 152°F for over a week, and above 145°F for another four days.[16] This turning also mixes surface material into the hot interior, helping to ensure that all parts of the pile get hot.

When my first pile was a few months old, I examined it with my nose. Our sense of smell is a direct connection to the microbial world, after all; an unpleasant smell often indicates something that can make us sick. The pile was about four feet by three feet by two feet high. It had been about five days since I'd dumped a bucket. There were no flies, and it smelled earthy.

With the pitchfork, I excavated down to the clay soil. The core of the pile, at several months old and well below the level of recently added deposits, looked and smelled like regular compost: dark black leaf bits, a rich earthy smell, and bugs scurrying around. There were no visible turd remnants, and no offensive smells— even with my nose a few inches away. I reached in and grabbed a handful of material from the bottom of the pile. It looked, felt, and smelled like dirt. Poop was transmuting into compost.

I can't overstress two key advantages of humanure: (1) it's simple and cheap, requiring only a bucket, a thermometer, a tiny bit of knowledge, and a square yard of land per several families; (2) it transmutes disease-causing poop into valuable, healthy soil. As far as technologies for global development go, few would benefit the world's poor more than humanure.

Globally, one in three people—2.4 billion people—don't have access to a decent place to poop. A billion of these people poop where they can—behind bushes, out the street. The rest poop first in buckets and then dump the buckets wherever, or else use "unimproved" latrines (a hole in the ground, or an outhouse on stilts with a hole in the floor).[17] A billion additional people have access only to "improved" pit latrines, which are themselves a major groundwater contamination risk which will worsen with increased flooding from climate change.[18] This picture is both inhumane and deadly, with contaminated drinking water directly causing an estimated 300,000 deaths per year and contributing to widespread disease and malnutrition, especially among young children.[19] In India, over half of the groundwater and 80% of the surface water is contaminated by sewage, and 40% of women who drop out of school in India say it's because there are no toilets in the schools.[20] Humanure composting provides a hygienic alternative. And because educated women have fewer children, humanure composting could even eventually help with overpopulation.

Humanure is safer than conventional waste treatment

The situation in the US, while better in some ways, is still far from perfect. In the US, 20% of households use septic tanks,[21] which collect household and human wastes. After a short settling time with minimal anaerobic decomposition, this untreated effluent, still with a huge pathogen load, is simply introduced into the soil (or "leach field"). Like pit latrines, these systems often silently cause groundwater contamination.[22]

The other 80% of waste in the US is piped to sewage treatment plants, which fail occasionally, and have difficulty coping with storm water. In Los Angeles, for example, our beaches are routinely closed after rains due to the dumping of large volumes of untreated sewage.

Even when all goes as planned, the process results in tonnes of toxic sludge. At the treatment plant, microbes are given only about two weeks to work at a temperature of approximately 95°F—not enough time and temperature to kill pathogens. The resulting "biosolids" are then sold to farmers and citizens. My dad used to put it on his lawn, and it smelled just like poop, which tells you something; also, the vast majority of biosolids are "Class B," meaning known to contain large pathogen loads.

Sewage treatment plants' input streams also include detergents, industrial waste, persistent pharmaceuticals, and heavy metals such as lead and cadmium, along with other toxins. Persistent pharmaceuticals and heavy metals build up in soils, and some crops readily absorb them.[23] Remediation necessitates removing the poisoned earth and trucking it "away." Personally, I'd not apply biosolids to my land.

We pay a high price for the convenience of flush toilets.

Taboo

In our culture, poop is taboo,[24] and for good reason: it can carry disease. But properly composted poop is no longer poop. It doesn't look or smell like poop. It wouldn't make you sick even if it did get into your food or water.

The power of taboo, however, lies in its emotional intensity. Evidence alone won't convince someone to relinquish a taboo. But time and familiarity will. Now that I've been composting humanure for several years, I can say that it quickly becomes not a big deal. Sharon tolerated this quirk of mine for the first three years. And then one day, she started using the leaf toilet, too.

One man's waste

Time races on. The planet warms, the corporatocracy consolidates, wars rage. And here I am, writing about poop in detail. How can I justify this use of my time?

I'm a mammal. I'm one man, tending my garden, coming back down to Earth. It takes time for good things to grow. I'm doing everything I can do within my particular situation, as fast as I can do it. And I'm old enough to know that patience is more effective than panic.

I'm suggesting that humanure composting is actually deeply relevant to our looming predicament. Apart from the ecological problems flush toilets cause, they're a powerful symbol of our mindless disconnection from nature. We go about our business pretending that we aren't animals, that technology has allowed us to transcend natural limitations. But no matter what we invent, we will always be nature.

Composting my own poop reminds me of what sustains me, and of my place in nature. It reminds me where I come from, and where I'm going.

Composting without poop

You might not feel ready for humanure composting. Fear not: just keep a basic compost pile, without poop. Maybe someday you'll try humanure. I also keep a worm bin, which is easy. It makes me happy to see the worms.

There are piles of information out there on both basic composting and worm composting. But I personally find humanure composting to be a much deeper practice.

Practical tips for humanure composting

If you decide to try humanure composting, here are some things I've learned which may help:

- Pick a shady spot for the pile.
- Use a compost thermometer.
- Keep a separate pitchfork for your pile. Don't use it for anything else.
- Kitchen and yard scraps can go into the pile. Contrary to composting dogma, I compost anything: meat, dairy, dead animals, pet poop. To keep animals away, enclose your pile in ½-inch hardware cloth (¼-inch if you want it mouse-proof).
- You can compost toilet paper, tissues, and paper napkins. Shredded paper and cardboard can be composted if added a little at a time.
- You can compost sticks, but break them up into small pieces first.
- Watch out for the plastic stickers on fruits and vegetables. It's truly uncanny how plastic never breaks down. Compost piles really demonstrate this.
- If your pile isn't heating up, you do need to find out why and correct the problem. Some possibilities: too much carbon, not enough nitrogen; not enough moisture; not enough oxygen. It is also possible your pile isn't big enough yet—heat seems to require a critical mass. Experiment and you will find the problem. After making a correction, it might take a day or two before the temperature changes.

Chickens

We keep hens and they delight us. My older son, Braird, is especially fond of them. After so many years of keeping chickens, I still love gathering their eggs from the nest box.

> I do not propose to write an ode to dejection, but to brag as lustily as chanticleer in the morning, standing on his roost, if only to wake my neighbors up.
>
> — HENRY DAVID THOREAU, *Walden*

Each hen lays a unique egg. We get brown, blue, red, and white eggs. Some are large and squat, some are long and skinny. Hens perform the alchemy of transforming kitchen scraps, garden weeds, and bugs into eggs. And they're delicious: organic "free range" store-bought eggs taste pale in comparison.

Feeding chickens

Most backyard chicken keepers feed their flocks on commercial grain feed, supplemented with ground oyster shells for calcium. Our chickens also eat kitchen scraps and forage for bugs from our yard.[25]

The problem of feeding a backyard flock in the suburbs was one of the things that led me to freeganism. Once, after shopping for groceries, I furtively checked the store's dumpster for something for the chickens. To my surprise, I saw an entire box of perfect eggplants. No good for the chickens, but great for a huge batch of *baba ghanoush*, which I shared with friends and neighbors. Thrown-out food—dairy, lettuce, a small amount of bread, and even occasionally meat—now makes up perhaps one-quarter of our hens' feed.

I also feed my hens grass and weeds in the spring. I've begun seeing the hens' favorite weeds, the lush grasses and dandelions, as a secondary crop. (Being happy to see "weeds" makes gardening that much more pleasant.) What the chickens don't eat becomes part of their bedding, all of which eventually finds its way onto my compost pile.

Like any being, plant, animal, or otherwise, chickens need a complete diet to thrive. Healthy chickens are active, their eggs have good strong shells, and their feathers are glossy and beautiful (unless they happen to be molting, in which case they're comically ugly).

Birth and death

Keeping chickens has changed how I think about life and death. Just like people, chickens are in flow, arising and soon passing away. However, chickens make this flow more apparent because it's more rapid, and because there's much less emotional attachment. But otherwise, we humans are cycled by nature *exactly* as the chickens are, arising, staying for some time, and then becoming something else.

As I said, our first two chickens were gifts from neighbor Ruben when he moved; Braird named them Ice Cream and Pie. Pie was a beautiful, protective Rhode Island red rooster. Roosters don't just crow at dawn; they crow whenever they damn well please. While I enjoyed the lively crow, the poor fellow who moved into Ruben's house was near to losing his mind. So of course I rehomed Pie.

This is at least partly cultural. Like Ruben, the neighbors on the other two sides of my house are Hispanic. They told me the rooster reminded them of their childhood; they liked it. But the new neighbor, while a kind person, asked me, angrily, "What sort of person has a rooster?!" Clearly, how we react to our neighbor's rooster depends on what we view as normal.

Hens continue laying fertile eggs for two or three weeks after the rooster leaves, so I borrowed an incubator and tried hatching a few. I chose eggs from our barred rock hen, Trophy. Barred rock hens and Rhode Island red roosters make sex-linked babies called black star: females hatch out entirely black, whereas males have a white mark on their heads. Given the business reality of hatcheries, where only females are valued, white is a death mark.

The eggs sat quietly in the incubator for 21 days. To develop, eggs *must* be kept at a steady 100°F. Fertilized eggs can't develop embryos if refrigerated. (That so many people believe that they can is a striking example of how disconnected we've become from our food.)

One morning, I found one of the eggs broken in half, next to a tiny bedraggled chick. She was black with no white spot, a girl. The other eggs didn't hatch—I'd probably waited too long to collect them after Pie left—so we bought her a brown puffball companion at the feed store. Braird named them Chocolate Chick and Butterscotch.

Babies need to be raised for three or four months until they're big enough to be introduced to the flock—a process requiring the establishment of a new pecking order. During this time, the two chicks lived in our kitchen, a chirping presence, and then in our garage. In December, integration of the chicks into the flock went smoothly. Then one night in January, the temperature dipped below freezing, a rare cold snap for Altadena.

There was a strange purity in finding Chocolate lying on her side on the floor of the coop the next morning, dead. Something cold and deep. I was shocked into the moment. It was a complete surprise, but then an instant later, it wasn't a surprise. For several more instants, the universe was me and her. The other hens rushed out for their breakfast around me as I remained crouched, completely still. I thought, how strange, how random, how natural.

The previous evening, I'd placed her on the floor of the coop. Afraid of the adults, she hadn't jumped to the roost and snuggled up like Butterscotch had. Instead, she went to sleep on the floor, and never woke up. Had there been a moment when she tipped suddenly over onto her side? Did any of the other hens notice? Braird, Zane, and I buried her deep within the compost pile.

Factory farming

As a society, we are radically disconnected from our food sources. We buy bloodless, commodified chicken legs in shrink-wrapped plastic, and we don't think about where the meat comes from. We don't want to know. Disconnecting from our empathy allows us to voluntarily—even eagerly—participate in a system that brutalizes other beings. (I believe that a similar disconnection allows us to wage war.)

When keeping a small backyard flock, you know that your hens are healthy and happy. But in factory farms, hundreds of thousands of birds are packed together under one roof. Hens live short and brutal lives in terrible conditions. Like other livestock within our industrial food production system, they are seen not as beings, but as raw material from which to extract wealth. Hens are killed after about a year of laying, when their egg output slows, and ground up for pet food. Such systems of wealth extraction are not aligned with the biosphere. Apart from being cruel, these systems are vulnerable to collapse.

Birds and other livestock are more susceptible to disease when they are crowded and stressed, so producers prop them up with antibiotics. Close quarters, poor health, and antibiotics make factory farms perfect breeding grounds for new strains of viruses and multi-resistant bacteria. And these new bacteria and viruses occasionally cross over to humans.[26] In 2015, a surprise outbreak of avian flu in the US killed 50 million commercial hens, turkeys, and ducks. A mountain of birds. The poultry industry is nervous for the next pandemic,[27] as are epidemiologists and public health experts—for far more ominous reasons.[28]

My own backyard flock has experienced collateral damage from the factory farming system in the form of Marek's disease. Marek's is a virus that manifests as tumors in the avian nervous system, causing paralysis of the legs and wings; there's no cure. Until recently, chicks were routinely vaccinated against Marek's, but the vaccine turned out to be "leaky": vaccinated birds, while remaining asymptomatic themselves, actively shed the virus. This may be weakening the gene pool and strengthening the virus.

I'm not against vaccines—those who are would do well to look at a few photographs of children with smallpox, and learn who Benjamin Jesty was, in order to better inform themselves

Practical tips for chicken keeping

Chicken keeping is relatively easy and rewarding. Here are a few tips:

- It's crucial to predator-proof the chickens' living quarters; humans aren't the only animals that enjoy a chicken dinner. To a raccoon, chicken wire might as well be thread.
- I built a run completely enclosed in galvanized hardware cloth dug down about a foot into the ground. This way I don't have to open a coop at dawn and close it at dusk. (Date nights and sleeping in are both good things.)
- I use deep bedding—about six inches of straw and garden waste—inside the run. Deep bedding eliminates poop smells, which can otherwise be a problem during rainy spells. About once a year, I cycle this bedding into the compost pile.
- I have a grain-feed and water setup that allows the hens to go four days without any human intervention, which is nice when we go on backpacking trips. A neighbor checks food and water and gathers eggs while we're out for longer stretches.
- Any food you give your chickens must be unobtainable to rats after dusk. I put the feed in a galvanized metal chicken feeder with a cover that opens only when a hen steps on a treadle. It's easier to prevent rats than to evict them once they've moved in, especially if you aren't willing to kill them.
- At some point, you may need to euthanize a chicken. I recommend using vinegar, baking soda, a five-gallon bucket, and compassion.

on the matter—but this particular leaky vaccine has turned out to be a curse for my flock. The breeders in my area are no longer vaccinating, but my fully grown hens are shedding the virus. While I've been forced to euthanize several pullets due to the disease (I won't be able to add new unvaccinated chicks until the vaccinated hens die off), my friends who hatch their own chicks and don't vaccinate have had no problems. If you start a flock, do so with unvaccinated birds.

Bees

The manzanitas are flowing. The scent of their nectar is divine as I lie underneath, feeling the sun's warmth and listening to the intense, comforting buzz of bees overhead. It's early February in the Sierra Nevada foothills, and I've been meditating in silence for seven days. My mind is so clear that it stays in each consecutive moment for long swaths of time. I feel perfectly at home in my body, on this Earth.

> Working with Bees is all about overcoming fear. The Hive is love incarnate. The Hive is the window to our new world.
>
> — SAM COMFORT

Going home, I drive Maeby through California's Central Valley. The almond bloom has just started, and I pass clusters of beehives by the side of the freeway. Beekeepers truck these hives across the country and even fly them in from Australia so that the almond industry can maximize profits.

Going backwards

I'm a backwards beekeeper: my bees come from local feral swarms, and I don't use any pesticides, antibiotics, or other treatments. Feral honeybees are a bit feistier than domesticated breeds, but they're not bad.

My first close encounter with a hive was in the cramped space beneath a trailer at the Santa Monica Fire Department's training grounds. I was performing a cutout: rehoming a feral hive that would otherwise be exterminated. Tens of thousands of airborne bees were buzzing angrily around me. The air smelled like bananas.

This scent was the bees' alarm pheromone, the signal to defend. *My instinct is to not be here.* I observed the urge to run, took a breath. My new bee suit seemed to be working.

This swarm was to become my first hive. I was working with two experienced beekeepers. Susan would cut a piece of comb from the bottom of the trailer with a knife and hand it to me, covered on both sides with bees and dripping with honey. I could feel bees vibrating eerily like tiny energized springs under my fingers as I gently held the comb in my gloved hands. I passed each piece up to Paul, who was standing outside and using rubber bands to fasten the combs into wooden frames which he placed side by side into two wooden boxes, stacked one on top of the other. This was major surgery for the hive, but better than extermination.

The hive at night

Two days later, I went back after dark to bring home those boxes. My first hive! I parked far from the bees. I put on my suit and walked over with a flashlight, paint stirrer, duct tape, and ratchet straps. After dark, all the foraging bees return and the hive is all home—the proper time to seal it up for a move. With shaky hands, I wedged the paint stirrer into the hive box opening and started securing it with duct tape. But I bumbled it up, the tape twisting and sticking to my clumsy gloves. Angry bees started pouring out like a liquid. Adrenaline surged through my body.

My heart pounding, I ran back to the car to regroup. I was unnerved by the hive, in the black night of this strange place with cars rushing by on the freeway just beyond the fence. But I had to accomplish my mission.

After a minute, I willed myself to walk back. Hundreds of bees were crawling around the entrance. I told myself "finish securing the exit, the suit will keep you safe," and I did. I told myself "put two ratchet straps around the hive boxes to hold them together," and I did. I told myself "carry the hive to the car," and I did. I drove home carefully, still in my bee suit, with angry bees buzzing around inside the car.

The swarm

An easier way to get a new hive is to collect a swarm. A swarm is a group of a few thousand bees, one of whom is a queen, searching for a new home. I recently collected a swarm in Altadena, from a friend's garden. I went at dusk so that the scout bees would be in for the night but I could still see. The bees were hanging from an artichoke leaf. I clipped away a few surrounding leaves and placed a cardboard box[29] under the swarm. Then I slowly clipped the leaf, gently lowered the swarm into the box, and taped it up.

In the heady days of spring, the nectar's flowing and new workers are hatching. If all is right with the hive, it may decide to swarm. The workers raise a new queen by feeding a regular larva a high-protein diet. And then, at some point, the old queen departs with several thousand of her daughters. They'll hang from a nearby branch for up to a few days while seasoned foragers scout for a new home.

The swarm's house hunt is one of nature's amazing stories.[30] While most of the bees in the swarm go into a powered-down mode to conserve energy, a few hundred scouts fly off in all directions. They travel miles seeking out potential hive sites: a hollow tree, an empty hive box, the rotting eaves of a house. When a scout finds a prospect, she'll fly inside and size up the interior, preferring a dark, watertight cavity with a small opening and a volume of around half a cubic meter. If she likes what she sees, she flies back to the swarm and excitedly communicates her discovery by doing a waggle dance on the backs of the quiescent bees. The dance tells the other scouts which direction and how far to fly, so they can see for themselves. If the scout repeats her dance many times, it means she *really* likes the place; more scouts feel her dance, so more scouts go to check it out. If they also like it, they'll come back and dance as well. Each scout bee gets to vote for the site by choosing whether to dance for it, and for how long.

Initially, there are many scouts dancing for different hive sites. But over time a consensus emerges. Once the decision is unanimous, they wake up the swarm by buzzing their wing muscles and everyone flies to the new site. The swarm's survival depends on the quality of their choice, and they rarely make a poor one. Their decision is egoless: a true beemocracy.

Back in my backyard, in the cool dark before dawn, I transferred my captured swarm into a small hive box with only five frames. I untaped the cardboard box and gave it a sharp rap on the ground to knock the bees down (they hang from the top). Immediately the sleeping bees erupted into a yowl of alarmed buzzing. In one motion, I opened the box and poured the bees into their new home. I was enveloped by the scent of bananas, but in the next moment I had the cover on.

The next day, I saw workers bringing pollen into the hive, brightly colored in gorgeous shades of red, orange, yellow, and blue. Bees use pollen to feed their young, so this means the queen is starting to lay and the hive plans to stay. The hive turned out to be strong and productive.[31]

Diversity and resilience

Unlike conventional beekeepers, I'm contributing diversity to the bee gene pool, and diversity means strength. So called package queens, shipped through the mail, are bred by only a few people in the US. Most of the 2.5 million commercial hives in the US—some 100 billion bees or so—are descendants of a handful of queens.[32] This poor genetic diversity and lack of local adaptation makes it easier for diseases and parasites to overcome hives. So conventional beekeepers employ an arsenal of chemicals, antibiotics, and fungicides to prop up their hives. Genetic strength declines further. It's a steady downward spiral.

Like our own bodies, a hive is a complex community comprised of an unknown number of symbiotic bacteria, fungi, and other life-forms. We know that bees depend on beneficial microbes to protect them from pathogenic microbes, and also to digest pollen, their protein source. What happens to the balance of microbes in a hive when a well-meaning beekeeper dumps in a cocktail of fungicides and antibiotics? We don't know yet, because we don't understand the complexity of hive biology. But we do know what happens to the pathogens: they develop resistance to antibiotics and fungicides.

To ignorantly meddle in this way with a robust and complex natural system is emblematic of our contemporary industrial mindset. In my experience, and in the experience of many other successful feral beekeepers, these interventions are unnecessary.

Natural systems find stability and resilience in diversity. This is true on every scale, from the inside of your intestine, to a meadow, to the entire biosphere. Farmers who plant in vast monocultures also attempt to create stability artificially by applying chemicals and using genetically modified seeds.

For crops that require bee pollination, monocropping also means that bees must be trucked or flown in. Bees can't coexist with monocultures: there's food for a few days while the crop's blooming, but for the rest of the year, there's nothing. In the middle of a field of almond trees in full bloom in California's Central Valley, you might not find a single bee if it weren't for commercial hives from far away.

So commercial operations wake up bees when there's still snow on the ground, feed them corn syrup and antibiotics, load them on semis and airplanes, and move them thousands of miles to pollinate almond trees laden with a cocktail of insecticides. Why make life so hard?

If I had an orchard

Unlike most other environmental problems, disappearing bees hit people at a gut level. Flowers with no bees seem uncannily sterile.

Colony collapse disorder (CCD) is a concern for commercial beekeepers and the industrial crops they pollinate. The cause of CCD could be stress from crop pesticides, beekeeping chemicals, corn syrup feed, poor genetic diversity, varroa mites (which may be encouraged by the use of artificial wax honeycomb), trucking bees around the country to commingle, pollen with ⅓ less protein due to rising CO_2,[33] or—since everything is connected—all of the above. These industrial practices weaken the bees' immune systems and upset the balance of hive microbes. Remnants of CCD colonies exhibit a large load of pathogens, demonstrating a weakened immune system.[34]

If I had an orchard, I'd plant different types of trees to maximize the blooming season, grow them organically, and intermingle the cropland with wild and weedy spaces. I'd keep beehives in place year-round. I might not earn as much, but I'd have happy, healthy bees. And this would make me happier than having a slightly larger number in some banker's database.

Not the honey

My greatest joy in beekeeping is not the honey. It's simply observing the bees. I used to keep two Langstroth hives under my kitchen window. I loved watching the bees' flight paths from the window in the morning sun, and the smell of nectar wafting into the kitchen.[35]

They are fascinating little animals, and only become more so. I watch them interacting with each other and going about their business in the hive, cleaning, building comb, or feeding brood. I watch the queen laying, surrounded by her retinue. Outside the hive, I see their tongues as they drink water. I watch them bringing back pollen, and I watch them foraging in flowers. If a bee lands on the back of my hand as I'm drinking my morning coffee, great— I can get a closer look.

Understandably enthusiastic

I used to think beekeeping was for everyone, but I no longer think this. The Backwards Beekeepers group in Los Angeles used to promote their work enthusiastically and welcome all comers. This led some folks to get bees who probably shouldn't have. These people liked the idea of having bees, but they were unable to get over their fear. As a result, they neglected their hives, which grew larger—and scarier. Experienced beekeepers had to intervene.

To "have" bees instead of to "keep" them—to allow a hive to grow without supervision, until it becomes crowded and grumpy—is dangerously irresponsible in an urban setting. Chained dogs and caged chickens can't escape from angry bees.

The honey

When a hive has a full top box (or "super") of honey and we're not heading into a dearth (a season without flowers; here, the hot months of summer), I'll take frames of capped honey, give them a sharp shake, brush the remaining bees off, and cut out the comb into a bucket. At home, I crush the comb, pour it through a paint strainer, and bottle it up. Honey from one's own hives is a joy. I process the leftover wax for candles. I give honey away.

A great gift

When I first started beekeeping, I had fear. But then I learned to observe my fear. The moment I'm able to look at my fear objectively, as an outside observer, I become calm and happy. In this way, bees give me a great gift: the opportunity to work on overcoming my fear of death.

I've learned to love these creatures who so improbably exist in this universe. Small creatures that spin sweet gold out of flowers in hollow trees—and protect it against all odds with tiny stingers? So many forces are arrayed against these beings. And yet, through it all, they persist, tirelessly pollinating flowers, bringing new life as they go.

Like the bees, we humans are also balanced at all times on the knife's edge of life and death. Like the bees, we persist. What honey do we make? We humans are impressive in our ability to tell stories and to make music. But the real honey we're capable of making is even better: compassionate love for all beings. Our job, in the brief instant we're granted on this lovely Earth, is to spin hatred, anger, and fear into pure love. When we love, we're being true to our nature, as when bees make honey.

CHAPTER 13

Opting Out
of a Broken System

May we look upon our treasures,
the furniture of our houses, and our garments,
and try whether the seeds of war
have nourishment in these our possessions.

—John Woolman (1720–1772)

Our predicament is the result of a vast industrial-commercial system of living, which can be viewed in various ways. It's the systematic fossil-fuelization of almost everything. It's the replacement of interpersonal transactions with money and debt. It's the redirection of distributed natural cycles with linear, centralized, monetizable flows of energy and resources—the stuff of billionaires' dreams.

It's as if humanity's cyclic connections to the land were cut by the scissors of the industrial system. We then plugged ourselves into the matrix, and we must now rely on that system for our survival.

Part of my response is to opt out of this destructive system. Opting out brings me the satisfaction of transitioning from consumer to producer. It can be playful, or delicious; sometimes it can be frightening; ultimately, it's fulfilling. Opting out is another form of reconnecting; as I lessen my dependence on global corporate systems, I naturally need to *opt in* to local biospheric systems.

Even as I explore how to opt out, I nonetheless remain deeply intertwined within the industrial system. After all, I use roads even when I ride a bike. But that's OK; this is a path of transition, a middle path, accessible to all.

This chapter is a smorgasbord of opting out of this system, a few of the experiments I've explored over several years. Of course, there are innumerable other ways to opt out; you may decide to go much further than I have. Cultivate stillness, listen, go where your principles lead you—and do what brings *you* satisfaction.

Opting out of stuff

So much of industrial civilization revolves around stuff, and much of this stuff quickly finds its way to the landfill. Does our stuff make us happier?

Good riddance

Getting rid of stuff frees up physical and psychic space—but getting rid of stuff can be harder than acquiring it. During an early summer heat wave, I decided to sell the window air conditioner in our bedroom which we hadn't turned on in four years. When I sat down to put it up for sale online, to my surprise I had second thoughts. I felt fear. What would we do in a heat wave? Without realizing it, I'd been mentally attached to that old air conditioner, a high-energy security blanket.

> The things that you own end up owning you.
>
> — TYLER DURDEN, in Chuck Palahniuk's *Fight Club*

I put up the listing, and someone came the next day. The window was freed up, the bedroom became more inviting. Since then, we've had some intense heat waves, but I haven't missed the air conditioner.[1]

My big old motorcycle took up lots of space in my garage. I didn't ride it enough to justify its existence; worse, I had a guilty sense that I should ride it more. Selling it freed up space in the garage, and in my mind.

Under our system of industrial capitalism, we've become hoarders, takers. As humans, we do need some stuff to survive;

but most of us now have more stuff than we need. I used to hoard stuff. Now I prefer getting rid of it.

Repair

Repairing things saves money, keeps stuff out of landfills, and keeps CO_2 out of the atmosphere. It also fulfills me at a deep human level. (It may be more of a man thing, though. I asked Sharon if repairing things makes her feel deeply human, and she said, "Not particularly.")

Early in the course of my adventures with my WVO car Maeby, I made a replacement aluminum handle for the pump I use to collect veggie oil.[2] Simple as it was, I showed it to my little sons proudly: "Made in America, by your dad."

To repair something, I need to project my mind into it. I need to see how its parts function and interact with each other. Only then can I come up with a good repair strategy. This process is a kind of flow; it brings me a sense of harmony with the physical world.

I've fixed leaky faucets, wobbly toilets, leaky roofs, the internet router, an old wheelbarrow the neighbors threw out, the solenoid in our dryer before we stopped using it. I've fixed a hundred things on Maeby, from her cruise control to her alternator, her front suspension to her rear axles, and seemingly everything in between. This is the price for keeping a 1984 car rocking and rolling.

It might seem like fixing things is a manifestation of attachment to them, but I find the opposite to be true. The ultimate fate of every fabricated item—every car or laptop or ceramic bowl—is to end up broken. During a repair, the transience of objects becomes so clear, and therefore my attachment to them becomes weaker. Things come, and things go.

Appropriate (low-energy) technology

We humans are technological animals, and technology will play a role in our response to our predicament. The most obvious example: renewable energy is replacing fossil fuels.

But there's an overlooked type of technology, maybe not so glamorous, that works with little or no fossil fuels. Things like push lawn mowers, solar ovens, treadle sewing machines, and bicycles—and techniques like orcharding, organic gardening, natural building, and plant medicine. These technologies are time-tested, relatively simple, easy to maintain, and ready to go.

I'm not suggesting that everyone start brewing beer or keeping bees, but I personally find using these kinds of low-energy technologies to be satisfying. And you don't necessarily need to set up a forge in your backyard; you can commission your local blacksmith to make those candleholders you need. A transition away from consumerism and toward local, people-based productive economies would require more of us to be producers. This could create meaningful jobs, result in tastier foods, build up communities, and make us happier in the long term.

Furniture

I build furniture when I need it, and my reason for doing so might surprise you: I often find it easier to build than to buy. Instead of spending a day going out to furniture stores, settling for something that's at least not awful, shelling out a small fortune, and figuring out transport, I prefer spending about the same amount of time creating a simple design and making the piece myself. I built precisely the bed frame we wanted (it wasn't available to purchase) and a small nightstand to go with it. This was a fun way to spend time with my sons, who worked the drill. I built precisely the coffee table we wanted, and precisely the kitchen storage shelves we needed.

Clothes

Sharon and I get cast-off clothes from friends and relatives, and from thrift stores. People discard clothes for various reasons, but usually not because they're worn out. It's easy to get decent clothes this way. (Jon Jondai, a master of opting out, articulates this point

Build a simple California king bed

Get the following pieces of wood: 6 pieces 4×4×15" (the legs); 14 pieces 1×4×72" (the slats); 2 pieces 2×4×66" (head and foot parts of the frame); 3 pieces 2×4×73" (right, center, and left parts of the frame). Also get some 1-inch screws, some 2.5-inch screws, and 24 3×⅜-inch lag bolts.

Lay out the frame on a flat surface, with the short pieces on the outside so that the frame dimensions end up being 66 by 76 inches. The third 73-inch piece is an optional center support for sturdiness. (If you leave it off, you will only need 4 legs.) Fasten the frame together with some 2.5-inch screws (drill small pilot holes). Then lay the 4×4 legs in the frame corners and fasten them to the frame with the lag bolts (two per contact face; use 3/16-inch pilot holes). Now you have an upside-down frame. Turn it over, lay the slats on with even spacing, and screw them to the frame.

To save time, I have the local lumber store cut my wood to size for a few dollars more. I use lag bolts for load-bearing joints, and star head screws for everything else. I try not to overbuild; here, the 4×4 legs are stouter than necessary, but I like the way they look.

Sharon and I agree that this is the most comfortable bed we've ever slept on. We use a basic five-inch-deep latex mattress. Try to find a mattress that hasn't been treated with toxic flame retardant. Or make your own.[3]

beautifully.[4]) Clothes seem precious to me, so I take care to make them last.

Clothes are a medium through which we differentiate economic class and group identity. So learning to embrace plain or hand-me-down clothes is a valuable practice for dissolving vanity and questioning repressive social structures. And the practice can go much deeper.

It's interesting to think about making clothes from scratch. You'd need to raise sheep, grow fiber, or hunt for hides; and then spin, weave, tan, and tailor. This thought experiment powerfully illustrates the impossibility of true self-sufficiency, and the necessity of community and specialization.

Of course, Gandhi famously made homespun clothes (*khadi*) a cornerstone of his practice; for Gandhi, khadi sat sweetly in the overlap between personal transformation, constructive program, and satyagraha. From my experience with homegrown food, I can imagine the immense satisfaction from reconnecting to the Earth in this way. I'd love to try it someday.

Electricity

On December 1, 2011, there was a powerful windstorm in Altadena that knocked down trees. Sharon was almost killed by a branch that fell in our driveway, seconds after she'd moved out of the way. Our electricity was out for three days.

Although my friends complained frequently about the outage, I loved it. The nights were quiet, and beautifully dark. We'd read for a while by candlelight and then go to bed early. I caught up on my sleep.

One of the main goals of *Being the Change* is to describe the experience of gradually opting out of fossil fuels while living an otherwise normal suburban existence. Leaving fossil fuels, in addition to reducing our emissions, holds the potential for radical reconnection and re-minding. It may even hold the seeds of a new society.

The Possibility Alliance is an intentional community in north-east Missouri, founded by Ethan and Sarah Hughes in 2008. They use no electricity and no fossil fuel. At night they burn homemade beeswax and tallow candles. And they don't view this as sacrifice—they love it. Adam Campbell, a resident, said, "There is no dogma in any of this. We're just having a great time living out a different way of being that we feel is a fundamentally better way to live, because it's more connected. It's more responsible. It's healthier. It's more vibrant. It's more participatory, and it's more fun. And visitors give us feedback that this is really true."[5]

My friends at Casa de Paz in Oakland, inspired by visitors from the Possibility Alliance, have also embraced the darkness; they light their nights with candles. After visiting them, I was similarly inspired; my family is trying it once per week. On "candle night," I do no work—no staring into the luminescent computer screen. Instead, Sharon and I play games with the boys. It's empowering to experience that *we don't need fossil fuel to live our lives.*

Earthen building

Over the course of three days at a local school garden, I had the good fortune to build a wood-fired earth oven alongside master adobe craftsman Kurt Gardella. We laid a foundation of gravel, built up the base with adobe bricks (formed onsite from the clay subsoil), and added a layer of beer bottles for insulation. The oven floor was made from bricks laid in sand. We formed the dome by covering a sculpted sand dome with cob (mud and straw plaster). When the cob was dry, we scooped out the sand.

> Most men appear never to have considered what a house is, and are actually though needlessly poor all their lives because they think that they must have such a one as their neighbors have.
>
> — HENRY DAVID THOREAU, *Walden*

Earthen building is the kind of work that makes you sleep well at night, a satisfying, meditative, and empowering experience: building on a human scale, with your hands.

While lifting my leg over barbed wire,
E. said, "You enter at your own peril."
I paused, and entered.

Now that I've built a simple oven, a modest earthen house feels doable. Earthen homes are works of art, planned with one's own mind and built with one's own hands, brimming with handmade personality connected integrally into the landscape. They stay cool in the summer and snug in the winter. Building them produces no greenhouse gases.

Someday I'll build one. I'd incorporate a rocket mass heater[6] into the design. Rocket mass heaters extract nearly 100% of the heat energy from their fuel, which can be "junk" biomass like pine cones and sticks. This means a fraction of the work chopping wood.

And earthen houses prevent debt. Jon Jondai built one in just three months, working two hours per day. Compared to someone with a thirty-year mortgage, Jondai had an extra "twenty-nine years and ten months of free time."[7]

Opting out of industrial food

Becoming more food-sufficient is a revolutionary act. In Chapter 12, I discussed tending fruit trees, vegetables, soil, chickens, and bees. In this section, I describe a few other fun ways I've opted out of the broken industrial food system.

Freeganism

I tell my mom that she taught me to not waste food so well that now I dumpster dive. It feels good to save valuable food from being dumped in the landfill, it feels good to get food without using money, and it's fun being part of the dumpster underground.

Two experienced divers showed me the ropes one Thanksgiving morning.[8] The streets were desolate at 7 AM, but I still managed to pull up ten minutes late. The two divers, E. and M., got out of a small white hatchback. I felt nervous, like I was showing up late to the revolution, but E. and M. greeted me warmly.

The dumpsters were in a barbed-wire enclosure festooned with "No Trespassing" signs, but the enclosure was set into a hill, and the back corner was easy to climb over. While lifting my leg over barbed wire, E. said, "You enter at your own peril." I paused, and entered.

Inside, E. nimbly vaulted over the side of a dumpster and then handed me a waxed box. "You never have to bring your own box," he said. He passed me a full white trash bag over the side of the bin. Inside, I found several cartons of eggs (each with ten or eleven perfect eggs), a large tub of foil-sealed yogurt with a cracked plastic top, some pita bread, dozens of potatoes, sweet potatoes, and orange peppers. There were bagels and cream cheese. I found some cucumbers and offered them to E. He asked, "Are they organic?"

Freegans tend to eat consciously, and E. and M. are no exceptions. They happily eat food from dumpsters—so long as it's organic. To put this another way, they'll eat food past some arbitrary expiration date, but they avoid food known to be literally laced with poisons (herbicides, insecticides, and fungicides).[9] I find their point of view to be reasonable.

E. and M. no longer buy groceries or chicken feed. Instead, they dive once a week. They keep what they need and donate the rest.

I now work with a local supermarket every two weeks, distributing their "spoils" to charitable groups and taking what my family needs.[10] In a country that wastes 40% of its food,[11] it feels good to live mainly off food saved from the dump, and to bring even more of this reclaimed food to others.

Freeganism, of course, is a stopgap, a way to live gracefully within a broken system. It reminds me that every bite of food is a gift. When I eat mindfully, with gratitude and with awareness of origins, I realize that food is a wonderful fuel that keeps my mind and body running well so that I can serve others.

Packaging

There's a flow of plastic bottles and plastic bags through my life. In our society, we love encasing our food in plastic. I'm amazed by how quickly the recycling bin fills up with plastic bottles and plastic produce trays. And our garbage is nearly 100% plastic bags, packed down into another, larger plastic garbage bag. If our technology is so great, why isn't our food packaging compostable?

I try to minimize my food packaging by growing food, and buying grains and legumes in bulk.

Vegetarianism

I don't eat meat. This works for me—I'm fortunate to have access to a variety of vegetarian foods—and it aligns with my principles of minimizing the harm I do to other beings, and minimizing my greenhouse gas emissions.

It's true that the industrial meat system is both inhumane and misaligned with the biosphere. And it's also true that eating meat with little or no awareness of its origin indicates the deep and widespread disconnectedness of modern life. Still, I don't believe that vegetarianism or veganism is unilaterally better than meat-eating. Generalization here is unhelpful. Would it make sense for the Inuit to be vegetarian? Has there ever been a tribe of vegan indigenous people? A resident of arid eastern Montana who respectfully hunts elk is eating biospherically.

> This searching for and picking wild food is one of the most pleasant, calming pastimes I know.
>
> — ROBYN DAVIDSON, *Tracks*

Jam

From time to time, I end up with big batches of homegrown or freegan fruit. So I make jam. A pound of fruit makes about a pint of jam. Yes, it does take sugar. I have friends who avoid sugar like the plague, but I think a little sugar is nice—especially when it's in jam. (It's a middle path, right?)

Making a batch of refrigerator jam (which skips the time- and energy-consuming sterilization step and must be kept in the fridge) takes me about an hour, counting prep and cleanup. For me, this making usually occurs shortly before midnight on Sunday

Wild foraging: Eating "weeds"

Some plants seem worth foraging to me, while others are too much work. It's a lifelong practice. Here are some of my favorite plants for foraging. Your region will have its own palette.

- **Acorns:** Acorn veggie burgers are worth the work of gathering, shelling, and leaching. As opposed to fast food, the ethos here is that the work makes the meal even more satisfying. I gather acorns from the coast live oaks along my bicycle commute between my house and the Jet Propulsion Laboratory. I use the flat of a cleaver to crack the shells, bearing down with straight arms.
- **Manzanita berries:** Manzanita shrubs grow in the San Gabriel mountains and in the foothills of the Sierra. The outer berry flesh has a delicious, unique flavor. Like acorns, manzanita berries were a staple food of the native Chumash people.
- **Stinging nettle:** If you pick the leaves from underneath and fold them in half, they won't sting your fingers or mouth. The leaves are rich in vitamins, minerals, and protein. They make a wonderful tea.
- **Wild mustard:** This plant grows profusely along the wooded stream bed near our house. It's considered an invasive in Southern California, but it's delicious.
- **Purslane:** A fast-growing weed, delicious raw and with more omega-3 than any other leafy veggie. I eat it as I weed it.
- **Miner's lettuce:** Miner's lettuce grows in shadier areas along the same stream bed as the wild mustard, but it's a native. It makes delicious salads.
- **Dandelion:** Dandelions grow as a weed in my yard. Some people love the greens raw, but I find them too bitter. They're delicious, however, sautéed with white beans, garlic, and lemon juice.
- **Arroyo willow:** I make the leaves into a pleasant earthy tea which is also a painkiller.

Dad's root beer

Brewing root beer is one of my sons' favorite projects. The idea is to make a dark and rich root tea, add sugar and yeast, and bottle. Adjusting the flavors to your taste is part of the fun. I bottle into used plastic seltzer bottles. (If you choose glass instead, be careful of exploding bottles.) The finished product is best opened slowly, outside. The sound after my first sip: mmm, ahhh.

If you can't find sassafras or are worried about the safrole it contains,[13] you can leave it out and bump up the wintergreen and sarsaparilla to compensate.

- 14 quarts water
- 8 cups sugar
- ½ c molasses
- ¾ c sarsaparilla root, chopped
- ½ c sassafras root, chopped
- 1 vanilla bean, chopped
- ⅛ tsp nutmeg
- 40 (or so) juniper berries, crushed
- ½ tsp salt
- ¾ c wintergreen leaves, chopped
- 1 pack brewer's yeast
- 7 star anise
- ½ cinnamon stick
- ¼ c raisins, chopped

Heat ½ the water in a large pot. Add everything except the wintergreen and yeast. Simmer for an hour or so; this is a good time to sterilize your bottles with an iodine-based sanitizer. Then add the wintergreen and the rest of the (cold) water. You can taste as you add the cold water to adjust the overall flavor strength and the sweetness. I like it strong—for slow sipping. Once it's at blood temperature, mix in the yeast, and bottle using a strainer. Store the bottles at room temperature; when they're rock hard (after three or four days), stick them in the fridge to stop the ferment. Makes about 14 quarts.

night, after I return home with freegan spoils. I might have a box with ten pounds of blackberries that aren't getting any younger, say. To wait would be to lose the race against mold.

But the next morning, the line of beautiful, full jars makes it all worthwhile.[12] Homemade jam is far better than what you can buy. And gifting it makes people very happy.

Jun *tea: Opting out of industrial beverages*

A friend dropped off an extra SCOBY (symbiotic culture of bacteria and yeast) one night. I'd never heard of kombucha, but the SCOBY looked so vile that I knew I had to try it. I brewed some green tea sweetened with raw honey, let it cool, and plopped in the SCOBY. After a few days, I screwed up my courage and took my first taste. It was still on the sweet side, with a tart kick—not bad!

Later, I started brewing *jun* tea, a fermented green tea beverage sweetened with honey that I like even more than plain kombucha. I sip bubbly *jun* instead of beer.

To make it, I use gallon pickle jars, with a rubber-banded cloth cover to keep out fruit flies. The SCOBY grows on top of the liquid; it's similar to a mother of vinegar. When it gets to a pleasant sourness, I put the *jun* in seltzer water bottles so that it builds up serious fizz. Sometimes I add flavorings like mint, ginger, pomegranate, or white sage.

Opting out of the money system

I still use money, but I do so carefully. Our money system encourages separation. Money leads to separation from our food, from our biosphere, and from each other. It depersonalizes interactions, transforming a human into "the checkout person." Money's addicting nature even leads to separation from our selves.

> Cultivate poverty like a garden herb, like sage.... Money is not required to buy one necessary of the soul.
>
> — THOREAU, *Walden*

If you really look, you'll realize that money is a collective delusion.

Opting out of wrong livelihood

I was once lost in the rat race's maze, a young man working on Wall Street as a computer programmer, dreaming, sadly, of money. After three years of this, I'd had more than enough of helping the rich get richer. I thought, "If I keep doing this, it will be a waste of my life. When I'm on my deathbed, I'll regret it." So I applied to graduate school in physics. Later, I realized I wanted work that more directly helped others, so in a warming world, I switched to Earth science.

It's important to continue examining our livelihood as we go through life. "Is my work meaningful? Does it make me happy? Does it help others?" In industrial society, we're conditioned to work in a peculiar way. The way we work differs markedly from the way people in other human societies have worked. Years of school give us a worldview, expectations, and a work ethic; then capitalistic culture pushes us to work long hours, often at repetitive, unsatisfying tasks. We become workaholics. We feel pressure to "keep up with the Joneses" and to "get ahead." We judge ourselves and each other by the size of our bank accounts. When we narrow ourselves like this, we miss opportunities to live a more satisfying life.

We can consciously opt out of work that harms ourselves or others. We can opt out of work that makes things that no one needs, or that feeds addictions, or that harms the biosphere. We can choose to leave jobs at liquor stores or weapons laboratories. Instead, we can seek meaningful work that helps others.

Leaving big banks

The heady days and nights of Occupy Los Angeles in 2011—and the many conversations I had with other protesters on the steps of LA's City Hall—raised my awareness of the problems endemic to our money system. During this time, I walked over to a nearby Bank of America branch and closed my accounts. This felt surprisingly good. I moved the money to a local credit union.

Next, I cut up my big bank credit cards. I had two, including one that earned frequent flier miles. If ever there was a symbol of

our predicament and its deep and interconnected roots in our social, technological, and money systems, it's got to be frequent flier
programs. The customer service rep in Gujarat transferred me to
a customer service rep in Florida who offered me layers of incen-
tives to stay. I finally persuaded her to close the account.[14]

Using cash

Switching to cash was Sharon's initiative. Says Sharon: "I like
sticking it to the credit cards. I know they're always lying in wait,
waiting for me to miss a payment by accident. You think you're
always going to pay on time, but they have actuaries, you know,
just waiting for the time your bank account number changes and
your direct payment breaks."

Paying with cash helps local businesses, as well. I called my
friend Jon over at Oh Happy Days, the vegan café and health food
store here in Altadena, and he said the fees he pays when customers
use credit cards amount to about 3% on average. He calls the credit
card industry an "interloper."

The Shopping Challenge

Sharon wanted to see if we could go 46 days[15] without buying anything, except for groceries (we do still buy some food) and urgent
maintenance items. I thought this was a capital idea. Many of the
habits we started during this Shopping Challenge became permanent because we enjoyed them.

To save fuel, I biked the kids to school in the bike trailer. To get
to UC Irvine, Sharon biked to the Metro, transferred to commuter
rail, and then biked to campus. On a typical weekday, we didn't
drive at all—a habit we've kept.

We adapted. We moved non-essential appointments that required driving. Buying lunch being off limits, I brought my lunch
to work, ate more fruit than usual, and didn't feel so heavy in the
afternoon. This habit has also become permanent—and saves us
thousands of dollars each year.

A surprising side effect of the Shopping Challenge was an

increase in the time we shared with friends—an increase in community. For example, on Day 2 we still thought it was important to get some gasoline; we'd forgotten to fill up, and the car was nearly empty. I called a friend, and he agreed to trade gas for eggs. We had his family over for dinner, and he and I siphoned some gas out of his car. Asking for help leads to connection.

Here's a complete list of what we bought besides groceries: bicycle brake pads ($10); co-pay for medicine (eye drops) for Braird ($13); six tomato seedlings ($15; save seed instead). We also paid our utility bills and mortgage. I biked more than usual—70 to 80 miles per week, up from my usual 50 to 60. The Shopping Challenge was great fun, and I highly recommend it.

Divesting from fossil fuels

The world will move away from fossil fuels sooner or later; investing in them seems increasingly unwise. Despite inevitable setbacks by ignorant people and politicians, the transition from fossil fuels will accelerate over time as global warming impacts become more severe and alternatives gain ground. Fossil fuel corporations will be forced to write off stranded reserves they'd long planned to sell; this will lower their valuations. In 2015, one of the world's largest banks, HSBC, advised clients to divest, saying those who don't "may one day be seen as late movers, on 'the wrong side of history.'"[16]

My own employer, Caltech, refuses to divest from fossil fuels at the time of writing. Caltech is not alone: most universities have so far refused to divest. I find it odd that institutions ostensibly serving humanity and young people would continue to support the industry responsible for global warming. Call me crazy, but it seems wrong to destroy the biosphere for a bit of profit.

Resisting war

The money system has many problems, but I especially dislike paying for war. About 45% of US income taxes are spent on the military.[17] While I don't mind paying taxes, I do mind that about half of what I pay funds institutionalized murder, which goes against my deepest principles. And in addition, the US military burns more fossil fuel than any other institution in the world. Veterans, I'm truly grateful for your individual service. But as an institution, nothing says "violently preserve an unjust global *status quo* that's rapidly destroying our biosphere" half so well as the US military.

In 2013, after talking it over—Sharon had reservations—we became war tax resisters. I took a few extra exemptions on my W2 form so that less federal income tax would be withheld from my paycheck.[18]

In April, I filed our 2013 federal taxes with a check for the balance of only 55% of our tax, along with a letter explaining my principles (which hopefully amused a bored IRS agent somewhere). I deposited the remainder into an escrow account run by Quakers,[19] to be released on condition that it not be used for war.

We understood that this was a symbolic action, and that the IRS would soon seize the rest. Most war tax resisters only have their bank accounts levied against what they owe, but some, randomly, are forced to serve years of jail time.

Facing off against systems of power is scary. In war tax resistance, we target the social system's lifeblood—money—and in doing so, we open ourselves to its full apparatus of legal, violent retribution. However, I have to say that standing up for my principles in this way makes me feel fully alive. More people ought to try it for that reason alone.

When we began receiving the threatening letters from the IRS, I held firm, confident that they'd soon seize the remaining funds and be done with it. But Sharon became increasingly nervous. Eventually she could no longer take the strain, so we paid up.

The US spends more on its military than the next ten nations combined.[20] This money causes great misery, largely for the sake of maintaining access to fossil fuels, and in my opinion could be better spent elsewhere—for example, transitioning to carbon-free energy. I also believe that the ongoing "war on terror" is making us *less* safe. In a global poll in 2013, 24% of the world's people said the United States is the greatest threat to world peace. The second-place country, Pakistan, had only 8%.[21]

Some will criticize me as "unpatriotic." However, for the above reasons, I believe that *supporting* the US addiction to war is unpatriotic. Supporting the US war machine is not unlike handing a drink to an alcoholic. Both acts lead to nothing but misery. War tax resistance is actually patriotism of the highest degree.

Money and the means of production

After college, Sharon spent two years teaching English in a small town in the Mongolian countryside. Traditionally, nomadic Mongolian herders lived far from cities, meeting their needs without money. Their animals gave them food, fuel, clothing, and shelter. But climate change is drying the Mongolian grasslands, forcing many herders to move to slums in the capital, Ulaanbaatar. Because they've been cut off from the means of production, they depend on money and are reduced to offering their labor for bare subsistence. The money system tends to replace a basic connection to the land, making people easier to exploit—slaves to money.

I'm not saying that we should all become yak herders. But I *am* saying that there's value in maintaining direct access to the means of production (such as land, resources, and information). The billionaires certainly realize this.

Leaving money

Once, when driving through Utah in Maeby, we stopped in Moab to meet Daniel Suelo. Suelo, whose name means "soil," has lived without money since 2001. He doesn't spend a dime, he doesn't barter, and he doesn't accept anything that isn't freely given. He spends many of his nights in a cave up in the hills. He dumpster dives, he hitchhikes, and he's one of the happiest people I've ever met. He's far happier than he was when he used money.

Your typical homeless person craves money. But Suelo isn't typical: his decision to leave money was clear-eyed and voluntary. Here's how he describes setting down his last $30 in a phone booth: "It was this total feeling of liberation, I felt like it was warm water pouring over my head, this total comfort, feeling like, wow, everywhere I go I'm at home, and everywhere I go I'm employed. The universe is my employer. It's like I'm always at home and I'm always employed. It's this intense feeling of security."[22]

I found Suelo in the Moab public library doing email, and after he finished, we walked over to a city park to share a meal. Sharon and I made tortilla pizzas with tomato paste, mozzarella, and olives cooked up on the camp stove, and Suelo pulled out some free-gan corn chips and canned salmon. He shared his food with us (the boys, who are not vegetarian, loved the canned fish), and we shared ours with him. Moab residents frequently stopped by to hug him and chat. Suelo freely gives a gift far more precious than money: he makes people happy.

I think Suelo is onto something real, and I can see myself following his path someday. But to live moneyless with small children would require a strong moneyless community; and there'd be no way to do Earth science, which I feel called to do. For the foreseeable future, I need to use money—but I don't need to love it.

Opting out and mindset

Opting out of the global consumerist complex is helping me come out of the old mindset I described in Chapter 6. The more I opt out, the better I feel, and the more I want to opt out.

Luminescent screens

The average American watches over five hours of TV per day—12 years over a 40-year period—and one-quarter of the content is advertisement.[23] The average child sees 25,600 TV ads in a year.[24]

THIS IS YOUR GOD.

—Message on dollar bills in John Carpenter's film, *They Live*

From an impressionable age, advertising conditions us to be docile consumers. TV makes us more passive and submissive to authority.[25]

Screens also take time away from other activities that are ultimately more satisfying and useful: developing skills, pursuing interests, getting exercise, engaging with others. Time for meditating can be found simply by watching less TV. (Trading TV for meditation is a good way to start thinking more clearly.) And deleting my Facebook account allowed me to focus my writing energy into this book. I opted out of Facebook in 2010 and haven't missed it once.

I had an eerie experience on the BART train in late 2016 in the Bay Area: nearly every person was staring at a screen. What's more, their expressions looked distinctly unhappy. It felt surreal, isolating, like something out of the twilight zone. Those little screens are addictive,[26] and they're making us depressed.[27]

Risk

I have a friend, a former spacecraft systems engineer at JPL, who welds as a hobby. I asked him to help me weld a trailer hitch for Maeby.

At first he responded with an enthusiastic yes. But an hour later, he called to say he'd changed his mind. He'd thought it over with his engineering hat on, and decided that the risk of highway failure wasn't worth the DIY satisfaction. While I see the wisdom in his decision, it raises an interesting philosophical point. In commercial society, we've engineered out a good deal of individual responsibility; this has percolated deep into our mindset. We use money to move risk into the corporate goods and services we purchase, which can be a very good thing (I'm actually glad most people don't weld their own trailer hitches). But have we taken it too far? Are there hidden costs?

Some of my actions do involve risk. Bees can be deadly. Humanure composting, if done incorrectly, can cause disease. A re-engineered old diesel running on WVO can stall on the freeway. When I take risks like these, I'm aware of them. I do my best to manage risk by becoming knowledgeable, avoiding temptations to cut corners, seeking expert help when necessary, and recognizing when I'm in over my head.

I think it's worth taking on some managed risk for the sake of building skills and increasing resilience. But I also take on this risk for the same reason I go surfing and skiing: it's fun.

Biospherism and the law

Consumer society has its preferred way of doing things; opting out of these preferred ways can bump against the legal system. Backyard beekeeping is illegal in Altadena (although this may soon change). War tax resistance is illegal. So is growing food in my front yard; growing food in vacant lots; using WVO as fuel; composting humanure; keeping chickens in my backyard; saving seeds; and using greywater from the washing machine.

But love is greater than law. Laws change and reflect the priorities of the dominant social group. And legal systems have been used as tools of the utmost violence and oppression. Indeed, the very foundations of the US were built on the legal genocide of native peoples, and the legal institution of slavery. Whenever love and the law are at odds, I'll choose love.

Still, despite this moral clarity, exploring mindful biospherism can be scary. In addition to occasionally pushing against the legal system, it pushes against social norms. I've faced ostracism from friends who don't want to think about global warming, and neighbors who want me to have a tidy green lawn. I've been on the "wrong" side of the police line at Occupy Los Angeles. Sharon was rattled for months after police banged on our front door for a honeybee complaint while I was out. And I feel I'm risking my career as an Earth scientist by writing this book.

In my opinion, this risk is all quite tame considering the stakes. I'd do more—if only I knew what. Global warming is a different kind of challenge than Civil Rights or Indian independence, because there's no clear oppressor. Rosa Parks could spark a revolution simply by refusing to vacate her seat on a bus. Gandhi could spark a revolution simply by burning his identity card. These acts took courage, but they were simple, and the course of action was clear.

I see no analogous nonviolent action for global warming; if I did, I'd do it.[28] We're all contributing to global warming. We choose to burn fossil fuels; no oppressive law forces us to do so. Therefore, the relevant "what to do"—stop burning fossil fuels— isn't civil disobedience. If I decide not to drive my car anymore, the cops aren't going to arrest me. It won't be on the evening news.

The fallacy of self-sufficiency

Opting out is gradually decreasing my dependence on the industrial system. The more we live outside of the industrial economic system—the more we take back the means of production—the more resilient we become. While there is ultimately no material security (this just isn't a guarantee the universe makes), opting out makes me feel more secure than hoarding money would. You can't eat numbers in a database.

However, opting out doesn't mean that I need other people less. It has actually increased my reliance on my local community, friends, neighbors, and relatives. The way I view community has changed. I used to view community, vaguely, as the people within a geographic boundary. Now I view community as the people I can count on. The reason I know I can count on them is because they can count on me. We are experienced in helping each other; we trust each other.

CHAPTER 14

Collective Action

*Climate change is a result of the
greatest market failure the world has seen.*

— Nicholas Stern

Changing myself allows me to explore and to model what life is like without fossil fuels. It also makes me happier. However, individuals changing themselves won't be enough. We also need collective action on a large scale if we hope to quickly reduce global carbon emissions and avoid increasingly catastrophic warming. As we change ourselves, we can simultaneously explore how best to use our growing awareness, empowerment, and connection to steer collective change.

The old approach

In 1997, the world's developed nations, with the notable exception of the United States, agreed to reduce their greenhouse gas emissions to 5% less than 1990 levels by 2012. The Kyoto Protocol was recognized at the time to be a modest pledge, not nearly enough to stave off dangerous warming. Even so, it was initially hailed as a great first step.[1] However, the US, China, and India never signed on, and as it turned out, no one took the agreement seriously. By 2013, global emissions were 60% higher than 1990 levels.[2]

The Paris Agreement of 2015 was another UN meeting, another set of promises. The good news is that, unlike the Kyoto Protocol, the Paris Agreement included the US, China, and India. The bad news is that the Paris promises are non-binding, susceptible to the whims of changing political leadership. Due to a variety of factors, but especially climate deniers in the US Congress, negotiating even the non-binding deal was so delicate as to be nearly impossible. Even in the unlikely event that these non-binding promises are kept, they won't be nearly enough to keep warming below 2°C.[3]

As we saw in Chapter 3, the annual global rate of CO_2 emissions has been growing exponentially at a rate of 2.2% per year (which means that every 32 years emissions have *doubled*). Our attempts to reign in global warming have so far failed spectacularly. What's more, if we keep trying what we've already tried—more empty diplomatic promises—we'll keep failing. Perhaps we should try something new.

A better approach: Carbon fee and dividend

Global warming is a market failure. Burning fossil fuels imposes huge costs on society that aren't included in the price of the fuels, primarily by causing global warming and respiratory illness. It's crucial to fix this market failure because few of us will voluntarily stop burning fossil fuels in a society that still strongly rewards this behavior.

This failure is at the systems level: information is missing, in this case price information. Without this critical information, it's impossible to engage a stabilizing negative feedback loop. Most economists therefore agree that the best way to address climate change is through a direct price on greenhouse gas emissions: a tax or fee.[4] This would fix the market failure by charging emitters—including you and me—for the "privilege" of disrupting the climate.

Carbon taxes or fees set a price on emissions. The price can be per tonne of CO_2 emitted, or better yet per tonne of CO_2-equivalents (to include other greenhouse gases and address methane fugitive emissions). This price is assessed far upstream, at the point of first sale—the mine, well, or port. It can be gradually increased, giving the economy time to adopt carbon-free alternatives while eventually making fossil fuels much too expensive to merely burn.

By definition, if a government keeps all or part of the revenue, it's a tax. But if the collected funds *are given back to the people*, it's a fee. This is not merely a semantic distinction; as we'll see, carbon taxes tend to act as an economic drag, while revenue-neutral fees tend to act as an economic boost due to increased consumer spending and job creation. And because taxes are never popular, a fee would be easier to pass into law than a tax.

A $30 per tonne CO_2e fee would increase the price of gas by 34 cents per gallon,[5] but the annual dividend would amount to $600 per person.[6] The less you emit, the more you'd come out ahead. Suppose that after seven years the fee had increased to $150 per tonne; this adds $1.69 per gallon of gas and yields a $3,000 annual dividend (or $12,000 for a household of four[7]). The size of the fee can be gradually adjusted to meet an emissions goal. The dividend should be sent to citizens as actual checks, not as a tax credit or a payroll adjustment. Checks would lend immediacy to the policy, increasing public support. I don't know about you, but I'd be quite happy paying $1.69 more per gallon of gas if the increase came with a $12,000 check each year.

This is a simple but far-reaching solution: it corrects the prices of *all* goods and services using fossil fuels. It would equitably address each of the seven categories of emissions discussed in Chapter 9. A carbon fee would make airplane tickets and natural gas increasingly expensive. It would make fossil-fueled electricity increasingly expensive, accelerating the transition to renewables. It would raise the price of gas and diesel, spurring fuel economy improvements, more carpooling, and more bicycling. It would make local and unprocessed foods increasingly affordable relative to their fossil-fuel-intensive alternatives; a similar systems dynamic would play out for goods and services. A carbon fee would even decrease landfill emissions by incentivizing biogas collection. Everyone would know that the price of fossil fuels would continue to increase steadily over time, so everyone, individuals, corporations, and institutions, would naturally transition equipment and investments.

An equal dividend would favor those people who use less fossil fuel. In practice, this would be progressive, since the wealthy—who tend to live in large houses and fly a lot—consume more fossil fuel than the poor; in the US, the richest 20% are responsible for 32% of emissions, whereas the poorest 20% are responsible for only 9% of emissions.[8] Over one-half of people would come out ahead by receiving more from the dividend than they spend on higher fuel costs.[9]

Instead of a dividend, revenues could be used to reduce existing taxes. However, such proposals are more complicated. They also tend to favor the wealthy, and could place an unsustainable burden on the working class. If revenue neutrality were implemented in this regressive way, a carbon fee would become increasingly unpopular over time.

A border adjustment on goods imported from countries without a commensurate carbon price would protect domestic businesses and prevent jobs from going abroad. Exporters to such countries would receive a refund of the fee. The adjustment would be based on the estimated carbon embodied in the goods.[10]

Cap-and-trade

Cap-and-trade is an alternative carbon pricing instrument in which total emissions are capped, polluters trade carbon permits (their right to emit a certain quantity of CO_2), and the market sets the price. The cap is then lowered each year.

Cap-and-trade sounds like a good idea: if your goal is to control emissions, why not just put a cap on them? But in practice, cap-and-trade schemes have serious problems, and the majority of economists prefer a carbon fee.[11]

First, cap-and-trade schemes are complicated and can be easily gamed. Emissions permits are traded on a market, but first the permits must be allocated to polluters, usually based on estimates of historical emissions. This perversely rewards the largest emitters, giving them an incentive to pollute during the period before allocation. Also, the murkiness and complexity of the estimation process could open the door to conflicts of interest, cheating, and corruption.

Second, cap-and-trade is expensive. It requires investment bankers, traders, and lawyers; they take a piece of the pie,[12] passing along higher energy prices to the people.

Third, these traders will naturally speculate and exploit the market (by means both legal and, at times, illegal). This adds to volatility in the price of fossil fuels, which undermines investment in renewables and efficiency and hurts low-income households.

Fourth, the dynamic of cap-and-trade perversely disincentivizes emissions reductions. Suppose some participants make great strides, and don't need to buy as many permits as supposed. This causes the price of permits to fall, which incentivizes others to pollute more. Furthermore, this inherently volatile dynamic is used by lobbyists to argue against other climate policies. The argument is that if some other policy effectively reduces emissions (e.g., by improving energy efficiency), permits will drop in value and the carbon exchange could collapse.

Some of these problems could be fixed by cap-and-auction. In cap-and-auction, carbon permits are directly auctioned to polluters. Since there's no trading, Wall Street speculation is eliminated. And there's no corruption-prone allocation stage. In my opinion, these are such major improvements that good-faith supporters of a cap should only ever consider cap-and-auction. But cap-and-auction doesn't solve the volatility problems, or how voluntary reductions undermine the prices of permits, leading—perversely—to disincentives to reduce.

Cap-and-trade did work for the far easier problem of ozone-destroying chlorofluorocarbons under the Montreal Protocol. But it has never worked for carbon. For example, the European Union emissions trading system (EU ETS) has experienced all of the problems I listed above, and has resulted in little or no reduction in emissions.[13] And international emissions trading under the Kyoto Protocol allowed a nation to emit beyond its allowance if it paid some other nation for some of its excess allowance. This sort of system encourages *more* fossil fuels to be burned, not less.

Those in favor of cap-and-trade often argue that it's more politically feasible than a carbon price.[14] This claim is dubious—a simple, fair, and easily understood carbon fee and dividend could well garner more public support than a complex cap-and-trade scheme—but even if it were true, it's a poor argument. A policy that doesn't work might be "more politically feasible," but it still doesn't work.

That said, it's possible to have a cap-and-trade scheme *and* a carbon fee, at the same time. But I suspect we'd be better off simply going with a carbon fee. Fighting the uphill battle to make cap-and-trade actually work will distract us, wasting additional time. And unlike cap-and-trade or cap-and-auction, we already have clear evidence that carbon fees work in the real world.

Carbon fees work

Lower emissions

British Columbia adopted an extremely modest revenue-neutral carbon fee in 2008. Revenue-neutrality was achieved by lowering taxes; BC now has the lowest personal income tax rate in Canada, as well as one of the lowest corporate tax rates in North America.[15] The carbon price started at CAD$10 per tonne CO_2 and increased gradually, reaching CAD$30 per tonne CO_2 in 2012 and remaining there (as of 2016). Even with this modest price, the *per capita* consumption of fossil fuels decreased by 17.4% between 2008 and 2012, whereas it increased by 1.5% in the rest of Canada.[16]

In a recent model study, a $10 national carbon fee and dividend (CFAD) starting in 2016 and increasing at $10 per year caused US emissions to drop to 69% of 1990 levels by 2025, and to 50% of 1990 levels by 2035.[17]

Economic benefits

The revenue-neutrality feature isn't really optional. As policymakers negotiate a carbon price, there will inevitably be some who'll want to spend the revenues on this or that program. This must be avoided, no matter how tempted environmentalists might be to fund this or that program. For if we aim to hit science-based emissions targets, as we should, higher fuel costs will crush low-income households—unless they get a full dividend.

Models clearly show that revenue neutrality is what creates the economic boost, by putting more money in the pockets of average people, which in turn creates jobs across multiple sectors.[18] If the model has the government spending the revenue, the economic boost is lost.

Real-world observations are consistent with these model results. Under its carbon fee, British Columbia's economy grew by 1.8% from 2008 to 2013 compared with 1.3% for the rest of Canada.[19] While this isn't strong evidence of an economic boost, it *is* strong evidence that the fee hasn't hurt the BC economy.

Given the reality of a political system that worships growth, it's good news that CFAD isn't an economic drag. Otherwise it wouldn't stand a chance.

Bipartisan support

In a rational world, CFAD would enjoy strong bipartisan support. Many conservatives do want to take action on climate change, but they prefer free market solutions without additional taxes or government regulation. CFAD fits their worldview extremely well, and perhaps they are slowly realizing this. In 2008, for example, the BC carbon fee was championed by the province's right-of-center party. And in the US, high-profile Republicans such as former congressman Bob Inglis, former secretaries of state George Shultz and James Baker, and Harvard economist Greg Mankiw[20] favor a revenue-neutral carbon fee. The libertarian Niskanen Center was founded in part to dissolve what is, in the Institute's view, a perverse conservative rejection of a revenue-neutral carbon fee.[21] To libertarians, a carbon fee is vastly preferable to additional regulations.

In 2016, a Florida representative, Republican Carlos Curbelo, with his Democrat counterpart Ted Deutch formed the Climate Solutions Caucus, a working group for exploring and advancing climate policies within the US House of Representatives. The caucus is bipartisan by design: every Democrat who wishes to join must come in with a Republican partner.[22] If the Republican party remains powerful, this might be a way forward for a carbon fee.

Hopefully, as more conservative leaders embrace CFAD, conservative voters will gradually recognize that it in fact aligns with their ideological worldview. This could change the game for Republican policymakers: denying climate reality would gradually become less rewarding. And liberals would be wise to agree to roll back some emissions regulations when negotiating CFAD, which would accomplish the goals of those regulations and then some.

Predictable and stable

A carbon fee locks in a higher price of carbon tomorrow, which makes investment in carbon-free alternatives an easier choice today. Recognizing this, six of the largest oil and gas companies are calling for the nations of the world to adopt carbon fees.[23] They realize that humanity will soon have to do something about global warming, and that of the possible options, a carbon fee provides the most predictable and stable framework.

In an open letter to the United Nations dated May 29 2015, they wrote, "We need governments across the world to provide us with clear, stable, long-term, ambitious policy frameworks. This would reduce uncertainty and help stimulate investments in the right low carbon technologies and the right resources at the right pace. We believe that a price on carbon should be a key element of these frameworks."[24]

Some green environmental groups see support from Big Oil as a mark against carbon fees. But this is woefully short-sighted. Humans will always need energy, just not fossil energy. If energy companies begin leaving fossil fuel and investing in the energy systems of tomorrow, that would be a wonderful thing. As economist Alan Blinder says, "I can hardly wait to witness the outpouring of ideas it would unleash."[25]

No regrets

If there's one thing that humans ought to agree on, it's that breathing clean air is a good thing. Even those who don't care about global warming would not regret a carbon fee.

Globally, air pollution from burning fossil fuels causes millions of deaths and costs trillions of dollars per year.[26] In the US alone, 200,000 people per year die prematurely from air pollution, mainly from road transportation and power generation.[27] CFAD would gradually fix this devastating problem, with no economic downside. Nations like India with severe air pollution could use CFAD to drive their transitions from dirty energy to renewables.

A pathway to international action

As mentioned above, implementation of CFAD would include border adjustments on goods imported from countries without a carbon price, or with a smaller carbon price. Suppose a nation without a carbon price—call it the dirty regime—wanted to sell a widget in the US market, and that this widget had embodied emissions worth $1 at the US carbon price. To get its widget across the US border, the dirty regime would need to pay a $1 tariff.

However, a US company making equivalent widgets now has an incentive to innovate and shift to renewables. The price of its widget falls, and the dirty widget loses market share. Since the US is the largest consumer market in the world, dirty regimes must also adopt carbon fees to remain competitive.

Coordinated international climate action has so far been an intractable problem. But CFAD provides an effective market-based pathway to international cooperation.

The role of individuals

At this point, a carbon fee may be our best chance for a livable planet. To get it, we citizens need to demand it. Ultimately, it's up to citizens to create the political space policymakers need in order to act. An amenable policymaker might move into even a small space, while a contrary policymaker will require a larger movement or even removal from office. But both the amenable and contrary policymaker require—and respond to—political space created by citizens.

So far, I've supported CFAD in three ways. First, I joined my local chapter of Citizens' Climate Lobby. CCL is an international group of volunteers fighting for revenue-neutral carbon fees. In the US, our primary goal is to create space for policymakers, especially Republicans, to support carbon pricing. We lobby our state and national representatives, and publish letters and articles advocating CFAD.[28] It's empowering to be part of a local community of advocates tied into a smart international organization. It makes me feel like I'm not in this alone: I have a team. Second, I give talks in my community. There are many opportunities to speak, and I'm sure this is true in your community as well. Third, I've written this chapter.

These advocacy actions bring me satisfaction. It's true that I'm only one voice, and I have little control over collective outcomes. But even while I wish I could do more, I'm doing all that I can—and contributing to something much larger than myself.

Only the first (real) step

Maybe, as you read this, the US has already enacted a national carbon fee. (One can hope.) But a carbon fee and dividend, by itself, won't be enough to transition to a biospheric mode of being on this Earth. A carbon fee would be the first real step we take to collectively address our predicament, but we'll need to keep walking.

As I discussed in Chapter 5, a sustainable human civilization will require us to reorganize around an economic system that doesn't depend on growth. We'll also need to find a way to maintain a stable population within the Earth's carrying capacity. One way or another, either through conscious planning or collapse, physics requires that our current growth-based system must change.

As individuals, perhaps the best we can do in service of this change is to begin telling the new story—the story of conscious biospherism—by living without fossil fuels and selectively opting out of the industrial consumerist system. As I've continued down this path, opportunities have opened to me, and I've increased my ability to change the *system* as well. There's a deep albeit nonlinear connection between individual and collective action. And there's no reason we can't actively engage at both levels.

I feel optimistic that nations will begin adopting sensible carbon pricing in the near future. But what other levers are there? And what comes next? Here, I'll brainstorm a few other ideas which in my opinion might help. Some are probably within reach. Others might be deemed politically impossible, but I mention them anyway in the spirit of planting seeds.

Use less energy

Electricity generation accounted for 37% of US energy-related CO_2 emissions in 2015.[29] If we're serious about mitigating global warming, we should pursue policies that significantly reduce our electricity use while at the same time expanding our carbon-free generation capacity as rapidly as possible.

In 2013, the US generated 4,070 terawatt-hours (TWh) of electricity.[30] Of this, 2,760 TWh were from fossil fuels, and the rest from carbon-free sources.[31] Based on my own electricity usage relative to the average US person's usage (see Chapter 9), I believe that the US could cut overall electricity consumption in half if we deployed sensible policies toward this goal.[32] This should be doable: total energy use *per capita* in the UK is less than half that of the US.[33] To be clear, I'm talking about more than increasing efficiency. I'm talking primarily about changing policies, norms, and behavior.

With halved consumption, we'd need to replace only 725 TWh per year of fossil fuels with carbon-free generation, instead of 2,760 TWh.[34] In other words, *we'd only need a ¼ of the carbon-free electricity infrastructure*, getting us to 100% carbon-free much more quickly.

And as we ramp up our carbon-free electrical capabilities, we can begin transitioning transportation and building heating and cooling sectors from liquid fuels and natural gas to electricity.

End fossil fuel subsidies

A fossil fuel subsidy is any government action that artificially increases the price-competitiveness of fossil fuels relative to renewables. Globally, fossil fuel subsidies amount to between $700 billion and $1 trillion.[35] Half of this is given to the fossil fuel industry as "consumer" subsidies—governments buying fuels such as gasoline and offering them below the market price. The other half is "producer" subsidies—governments essentially giving away land and water, government-sponsored research and exploration, and favorable government loans. I've not included the military cost of protecting fossil fuel assets. I've also not included real but externalized costs due to respiratory illness and global warming (estimated at $4 trillion in 2015)[36] which aren't government actions—and can be best addressed by a carbon fee.

Of course, Big Fossil fights hard for its subsidies, which increase profits and hold back renewables. But as taxpayers, we're giving handouts to the wealthiest industry in the history of the world at a time when we desperately need to move in the opposite direction. This puts international agreements on climate change in a sobering context: if our governments continue unfairly boosting the fossil fuel industry, how committed to climate action can they really be?

Fight in the courts

Kids are suing the US government on the grounds that the defendants have a constitutional obligation to protect the atmosphere for future generations.[37] The suit was brought in 2015, when the 21 youth plaintiffs were between the ages of 8 and 19 years old. The children are advised by lawyers from a group called Our Children's Trust and climate scientist James Hansen, who is also a plaintiff.

A few weeks after the kids brought their suit, the fossil fuel industry joined the government in the fight against them. Trade groups representing the industry moved to intervene on the grounds that the lawsuit would "impair the interests" of corporations such as ExxonMobil, BP, Shell, and Koch Industries.[38]

A key part of the plaintiffs' case is that the government knew it was harming them, and continued doing so deliberately. Indeed, the US government has known about the danger of climate change for over 60 years. For example, in 1955, the US Office of Naval Research linked burning fossil fuels to rising atmospheric CO_2 levels, rising temperatures, and increasing hurricane frequency; and in 1979, a report from the National Academy of Sciences warned that a "wait-and-see policy may mean waiting until it is too late."[39] Viewed in this light, it's shocking that the government would choose to support, subsidize, and accelerate the burning of fossil fuels despite clear knowledge of the danger.

The kids' lawsuit is far from frivolous. In 2015, a Dutch court ruled that reducing greenhouse gases was a state obligation.[40] And two US judges have ruled that the plaintiffs have standing. President Obama and the fossil fuel intervenors moved to dismiss the case, but US Magistrate Judge Thomas Coffin ordered their motion be denied. Then, on November 10, 2016, Federal District Court Judge Ann Aiken not only affirmed the order, but completely rejected the defendants' arguments to dismiss. The case will likely go to trial in 2017.

If the youth's groundbreaking fight in the courts is successful, the US government will be legally bound to reduce greenhouse gas emissions, forcing Congress to search for an effective and economically palatable means of doing this. This could lead to CFAD.

Redistribute wealth

If we are to have some chance of getting to a steady-state economy, we'll need to rethink wealth and power.

For example, how much money does one human need to live well? Does any single mammal really need $100 billion? Is it OK for the wealthiest 62 people to own more than the poorest 3.6 *billion* people?[41] What would the world be like if this hoarded wealth were distributed equitably? The monopolization of resources far beyond what an individual actually needs happens nowhere else in the biosphere. It's anti-biospheric.

Where does all that billionaire wealth come from? I hold that ultimately there are only two places it *can* come from: owning and extracting the Earth's natural resources, and owning and extracting labor (in the case of billionaires, the labor of others).[42] Over the last several decades, unions have been dismantled,[43] and productivity gains from robots, computers, and other technologies have accrued disproportionately to the rich.[44] These technological innovations tend toward extracting wealth from natural resources more efficiently while eliminating the need for human labor altogether.

It's unconscionable to have a society with both multibillionaires and workers so impoverished they can't even afford basic health care. Somehow we need to find the means to redistribute this wealth, and engineer checks and balances to keep it from accumulating in the hands of a few.

Get money out of politics

Allowing unlimited wealth accrual also dangerously distorts society. As we discussed in Chapter 5, concentrated wealth influences the power structure to concentrate yet more wealth in a runaway cycle.

When I started working on this book in 2012, getting money out of politics was a fringe idea held by a handful of Occupiers. The Occupiers planted that seed; it has since grown into a mainstream discussion. Corporate control of politicians has been a huge barrier to action on global warming. We need strict new campaign finance laws: corporations must not be allowed to contribute to political campaigns, and neither should any sort of organization. We need a system in which politicians aren't expected to or allowed to accept money from donors. And corporate lobbying should quite simply be illegal.[45]

Rethink global trade

To promote economic growth after World War II, the nations of the world, led by the US, created the World Bank, the International

Monetary Fund (IMF), and eventually the World Trade Organization (WTO). These institutions broker international trade deals that place growth above all else; legally overturn environmental and social protections that are seen as "barriers to trade"; saddle developing countries with crippling debt; and function as wealth pumps from the poor to the rich—all under ostensibly noble goals such as improving the quality of life in developing countries.[46]

Donella Meadows wrote that global trade "is a system with rules designed by corporations, run by corporations, for the benefit of corporations. Its rules exclude almost any feedback from any other sector of society."[47] These deals help corporations advance their goal of growth at all costs, but are devastating to the global poor and to the biosphere.

While there's a place for international trade—especially for climate-aware international trade under a carbon fee system— there's certainly also a place for strong local economies. I'd suggest that we ought to look first to local sources for our needs. In this way, we can take care of our communities, which take care of us in turn. Trade should primarily benefit people, not corporations.

Toward a stable population
We also need policies that encourage an end to population growth. I believe that contraceptives and age-appropriate sex education should be readily accessible to every human being on the planet, male or female, no matter how poor. Globally, 40% of pregnancies are unintended; not only does this contribute to population growth, it's a huge public health challenge.[48]

I also believe that all humans, including girls, should be guaranteed an education through high school, and that equal rights should be afforded across the board. Not only is this the fair thing to do, but educated and empowered women have fewer babies. Unfortunately, these policies face stiff resistance from conservatives, as well as from major world religions.[49]

Ultimately, we should consider the relationship between food production and population, the root cause of population growth.

For example, what if we adopted policies whereby agricultural land was gradually taken *out* of production and given back to the non-human ("wild") part of our biosphere? This would reduce global food production, leading to less meat consumption and perhaps to a tiered pricing structure for food whereby the global poor paid less than the global rich.

This may sound crazy, but is it really any crazier than causing a sixth mass extinction and deeply reducing biodiversity, arguably the best measure of the biosphere's health, for the next ten million years?

Half-Earth

To take this last idea a step further, what if we transition to a world where half the Earth's surface was set aside for nonhumans? Edward Wilson, one of the world's preeminent biologists and an expert on the needs of the biosphere, proposes exactly this in his book *Half-Earth*.[50] Wilson argues for reserving half the Earth in order to sustain biodiversity, which depends on the land area afforded it. (As a mathematical rule, the larger the island, the more biodiversity it can have.)[51]

The news these days is depressingly full of stories about critically endangered species. Often these stories center around heroic efforts by conservationists, efforts that, while laudable, are too little, too late. Extinction is accelerating exponentially, and the mindset of industrial society is blind to meaningful solutions that address the underlying problem. Instead, society gravitates toward speculative techno-fixes, like banking DNA. I worry that this is magical thinking.

As a policy goal, half-Earth seems breathtakingly audacious on a planet so teeming with humans. But just a few generations ago—not even a blink of geologic time—it would have seemed equally breathtaking to imagine that our species could so dominate this planet. This half-Earth goal transcends tribe and nation, and its implementation would require careful planning and cooperation to ensure fairness. Half-Earth would address the "jobs or nonhuman world" false dichotomy at a global scale, and as such interconnect deeply with population growth and equitable wealth distribution. But it's not impossible, and as humans increasingly experience our dependence on the biosphere, we will come to clearly see that half-Earth isn't just for nonhumans—*it's for us*. We're all Earthlings.

International cooperation

People often say that global warming requires a "World War II effort." But we actually need more than that level of cooperation: we need *global* cooperation. After all, during the war, half the world was fighting the other half. Humanity needs to find a way to set aside tribalism and come together to address *global* warming and overpopulation.

It's quite remarkable that the US, a country that prides itself as a shining beacon of freedom and enlightened rationality, is pulling the world down into the depths of climate disaster—while authoritarian China has seized the reigns of global leadership and is doing what it can.[52] I'd love to see my beloved country step up and lead, or at least pitch in a little. But it seems as though we're experiencing a glitch in democracy, if not something worse: as a nation, we can't even agree on basic facts. Perhaps democracy has been sickened by Facebook and Twitter. I have a new appreciation for its fragility.

In the long term, humanity will need to transcend competitive, murderous nationalism. That's the evolutionary challenge before us, and it will require humans to stop viewing life as a zero-sum game, to stop being so afraid. In the short term, though, the most promising path to real international cooperation might be through a system of carbon fees.

Tipping points

Collective change sometimes happens rapidly. While the precise connection between individual and collective change remains unknown, it's nonetheless clear that collective change can (and does) occur unexpectedly.

To fully address our broader predicament, we need a *paradigm shift*—within the social system that now comprises most of humanity—from growth, consumption, and separation to sustainability, living well, and connection. Before this can happen collectively, some fraction of individuals will first need to make this shift within themselves. However, in my experience, the illusion of separation is so strong that it can only be dispelled through a practice, such as meditation, that strengthens equanimity and weakens the ego.

Meanwhile, CFAD is a useful policy stopgap because it can begin ramping down emissions even within the *existing* paradigm. And individuals can begin exploring life without fossil fuels even without a meditation practice. Moving away from fossil fuels will lead us to slow down and question existing systems, taking us in the direction of broader shifts.

Being a scientist-advocate

In this chapter, I've presented my opinions as a *human*, not as a scientist. Although my job is to do science, as a human I have as much right to respectfully express my opinions as anyone else. I've done so here in order to balance my emphasis on individual action and emissions reduction elsewhere in the book.

And perhaps scientists even have a *responsibility* to advocate for acceptance and action. Because scientists are in a position to know certain things first. We're the ones face-to-face with global warming and its impacts on a daily basis: we know how serious it is. When scientific knowledge demands action, no one is in a better position to speak out than scientists.

When I feel unsure about whether or not I should speak out, I think of the billions of people with no voice on the matter. I think of those who are most vulnerable. I think of my children. And then the decision to speak out is easy.

CHAPTER 15

Community

Pshaw! What are neighbors for
but to help each other out?

— Mrs. Scott, *Little House on the Priarie*

What's life all about? Is it about hoarding material riches? Or is it about eating well and spending time with happy people? Community is an essential part of being human, so much so that I find it hard to imagine life without it.

The jungle

Every animal has unique attributes for thriving in the wilderness. The deer has speed and agility. The bear has powerful limbs and a keen sense of smell. The hawk has mighty wings and sharp eyesight.

I used to think that humanity's unique attribute was a big brain. We survive by our wits and our technology, after all. But this is only part of the picture. Our brain is merely a prerequisite for the actual attribute that allows us to thrive: community.

Imagine being deep in the jungle by yourself, naked, with no tools. Even with your big brain, you probably wouldn't survive. To thrive and be happy would be even more difficult. Your big brain isn't enough by itself.

With tools and gear, you'd fare better. But then, in a sense, you'd no longer be alone. Tools and gear represent community condensed into material form. They were designed and perfected

Without community
we are like blind hawks.

by many humans exchanging information, learning from one another's mistakes, building on one another's innovations over time. And with skills you'd fare even better. Like your tools, the skills are a form of community, condensed into knowledge and neuro-muscular technique. They were also developed by a community of humans, refined over time, and passed down.

Now imagine you're part of a tribe. It's night. You hear millions of insects chirping, the sounds of night birds, the lion's roar. You feel safe among your close friends and relatives in the clearing. A child cries in a nearby dwelling and is comforted. Your survival depends on this community. In community, you learned how to find good food, how to make a snug dwelling, how to use plants to stay healthy. You don't take your community for granted. Gifts are exchanged freely, smiles and laughter are the norm, and there are many celebrations. These things keep the community strong. They're expressions of gratitude.

We are community animals. Without community we're like blind hawks. I'm concerned that our communities have atrophied and been largely replaced by some corporate facsimile.

Simple, place-appropriate living

Henry Thoreau, writer and naturalist, was fascinated by the indigenous people who had inhabited his region before the white colonists. Inspired by them, he experimented with simple living, seeking to free himself from the economic fetters of contemporary society. Thoreau was impressed that Penobscot lodges were "constructed in a day or two at most" and were nonetheless "as warm as the best English houses." By contrast, Thoreau noted, it took "from ten to fifteen years of the laborer's life" to pay off the cost of an average house in his day.[1]

Thoreau concluded that his contemporaries led a more difficult life than the indigenous people who preceded them. These original people had the advantage of a mature community intimately adapted to their specific bioregion. Like many—perhaps all—indigenous peoples, the Penobscot passed down their place-appropriate knowledge and skills from generation to generation. By contrast, our modern way of life literally steamrolls over bioregional differences: US subdivisions are strikingly similar, from Houston to Anchorage. We've lost the bioregional knowledge of our elders. Instead, we depend on globalized commerce. And if anything, we've become still more economically fettered than Thoreau's contemporaries: instead of 10 or 15 years, we typically need 30 years to pay off the cost of an average house in the US.

Fabricated suburbs are a luxury of the age of fossil fuels, in that fossil fuels provide the *power* to steamroll bioregional particulars. Instead of building place-appropriate dwellings, we crank up the AC or the heater as desired. Instead of building human-scaled, integrated communities, we drive heavy metal and plastic boxes called "cars" long distances on congested "freeways." And, of course, we ship in food from afar, food with no connection to the local ecology.

As we transition away from fossil fuels, we'd do well to learn from the place-adapted communities that preceded us—at least to the extent possible, considering that we've so thoroughly erased them. Strong communities adapted to place help people live well and live efficiently, because living adapted to place simply takes less effort and energy. If you think about it, something's out of whack when it takes 30 years to pay off a simple dwelling.

A biodiversity of cultures gives us the building blocks we need to imagine and create a new story. Unfortunately, as Donella Meadows wrote, "People appreciate the evolutionary potential of cultures even less than they understand the potential of every genetic variation in ground squirrels."[2]

Asking and giving

Low-energy living requires me to ask for help from others. There's a lot to get done, so on a more-or-less daily basis, I ask for favors (and also let a lot of things remain undone). For example:

• A friend and I had planned to meet at our shared community garden plot, but it turned out that I didn't have the time or the energy to bike up there. I asked him to bring me some grapefruit from the tree in the garden. He ended up staying for dinner, and we played some music afterwards.

- We have a network of friends with kids with whom we swap babysitting duties.
- I asked a friend who's a mosaic artist if she'd teach the boys and me how to mosaic the perimeters of broken concrete raised beds in our front yard. She ended up staying for dinner, as well.
- I found that I had no seeds to start zucchini when the time was ripe, so I asked my neighbor for some.

My friends also ask me for favors, and I'm eager to help when I can. I'm actually glad when they ask me for help. Interestingly, it feels like everyone comes out ahead. The total benefit that occurs within the community is greater than the sum of the parts. A practical explanation for this is that we tend to seek help when we are in need, and we tend to give help when we have plenty to spare. So when we receive help, we're grateful, it's of great value; and when we give help, it's rather easy to do. I call this *asymmetric economy*. This is how the biosphere works: a bird eats a berry, poops out a seed, a new berry bush grows somewhere else. But there's also a human reason why asymmetric economy works: when we help someone with an open spirit, we become at least as happy as the person we help.

Seeking help, giving freely, and saying no (when it's necessary to say no) are three daily practices of community.

Community awareness, within biking distance

Some of you may think that interacting with neighbors in this way is the most natural thing in the world, even silly to write about. But community doesn't seem to happen automatically in 21st-century America. I lived most of my life without this kind of interaction. I found that I had to *practice* basic community, and that it only gradually began to feel natural.

As is typical in the US, in my neighborhood, people aren't outside much, and they watch a lot of television. To engage with existing community structures, I first had to find them—I had to develop a sense of community awareness. This involved gradually getting to know a lot of like-minded people within an easy bike ride, becoming aware of what their interests are and how *they* interact with the community, and developing a sense of what the community needs.

As I got to know people and establish trust, they'd introduce me to their friends. In this way, I gradually tied into existing networks. I've become one thread in a fabric, one node in a network. I'm linked to other nodes—the people in the community who trust and respect me, and whom I trust and respect—and I influence them as they influence me. Together we decide where to invest our time and energy. It's an organic, inexact structure.

In an earlier phase of my experiments in community, I tried connecting with people via Facebook. But I found this unsatisfying.[3] I personally prefer face-to-face interaction.

Community networks

Natural systems, from the soil to our brains to human communities, draw their power from networks. When there's a crisis, our network pulls us through. When there's no crisis, our network lets us thrive. The depth and redundancy in the network translates to resilience. Below, I list a few that have been especially important for me, along my own path.

Citizens' Climate Lobby

Citizens' Climate Lobby (CCL) is a grassroots organization of volunteers working toward the carbon fee and dividend described in Chapter 14. Our local chapter has 30 or 40 active members. These folks were my antidote to feeling crazy for being concerned about global warming, several years ago when it seemed like no one else cared.

Backyard produce exchange

Altadena is a small unincorporated community nestled against the San Gabriel Mountains in northeastern Los Angeles. Many Altadenans grow fruits, vegetables, and herbs in our yards. Backyard chickens and bees are commonplace; there are even backyard goats.

A group of us meets every few weeks at a local park to exchange excess homegrown food. If my lemon tree is producing more than my family can eat, I'll fill a bag with fruit, head to the crop swap, and come home with a grocery bag full of homegrown food that others had in excess.

We call ourselves RIPE (residential in-season produce exchange). Crop swapping is a miracle of gifting in which everyone trades something of little value (excess produce) for something of much greater value (a variety of homegrown food, vegetable starts, etc.). In other words, crop swapping leverages community to create value, tangibly—another example of an asymmetric economy.

We also use a mailing list to share gardening knowledge, organize bare root fruit tree orders, offer unwanted stuff, and so on. I recently surveyed RIPE members to find which fruit trees have been successful and which haven't, collated this data, and then shared this with the group. This Great Altadena Fruit Project harnessed local knowledge in a useful way.

I've also used RIPE as an in-person knowledge exchange. I learned how to prune by organizing a moneyless pruning class at my house (an experienced pruner volunteered to teach it). I also designed a simple laundry-to-landscape greywater system,[4] and I turned my installation into another moneyless class. I benefited from the help, and my helpers gained knowledge and experience.

Sangha (spiritual community)

I find it essential to have a spiritual community, connection to people on a similar spiritual path. This community, or sangha, inspires me and practices with me. It helps me keep my practice strong.

Once per week, Sharon and I invite local meditators to our house for a group sit. This is a great way for us to charge up our vipassana batteries.

We also attend and help to organize "Awakin Circles," informal gatherings that start with an hour of receptive silence (from any or no tradition), continue with a circle of sharing and deep listening, and end with a simple vegetarian meal.[5] We experience that all are our friends, and none are our enemies. It's really extraordinary to see this transformation happening in real time.

Serving at meditation centers (cooking, cleaning, and washing dishes for the meditators) also connects me with my sangha.

There is help and love everywhere, if only we are able to sense it. Trees are a part of my sangha, and sunlight and rain, and the plants in my garden, and the bugs in the soil, and certain places in the mountains. I find refuge in these things.

Community garden

Even though I have a garden at my house, I enjoy being a member of the community garden, which is a mile away up the hill. There's great gardening energy there, with dozens of experienced gardeners with whom to trade ideas, produce, and plants. Gardening is a strangely effective community builder. I think we humans might instinctively associate gardening and community.

Transition Towns

I have a dozen friends who meet regularly under the aegis of the Transition Towns movement, an international network of local groups exploring resilient alternatives to fossil-fueled industrial life.[6] This group of passionate activists and doers, Transition Pasadena, organizes events with the intention of building local community. Whereas CCL is sharply focused, Transition Pasadena covers a broad spectrum, pushing for such things as free food gardens, bicycle infrastructure, local art, a Styrofoam™ ban, and linking up mulch-providing tree trimmers with mulch-needing homeowners. My favorite Transition Pasadena project, though, is the repair café: people bring in their broken things and volunteer tinkers and tailors mend them for free.

Church communities

Having reduced my own carbon emissions, I'm now interested in helping to pioneer similar reductions within a group. Three church communities in my neighborhood have expressed interest in reducing their emissions together, in community (two are Unitarian Universalist, and the third is Episcopal). Church groups around the world, called to actively protect creation for future generations, could be a powerful force for grassroots climate action. They can begin by quantifiably transitioning away from fossil fuels. From what I've seen so far, this can be done with a spirit of creativity, subversiveness, sacredness, and fun.

Community solar

Instead of putting solar panels on my roof, I decided to do what I could to push for solar panels for all of Los Angeles County. This way I'll be able to get 100% renewably generated electricity for cheaper than rooftop solar—while simultaneously helping to do the same for ten million other people.

The mechanism behind this is called community choice aggregation (CCA). In California and a few other US states, communities are legally able to band together to form an alternative electric utility.[7] In doing so, they can choose exactly how their electricity is generated. The existing utility company is required to deliver the electricity to residents (and can charge a fair delivery fee). By installing local renewable generation, CCAs also invest in their own communities, creating local jobs and infrastructure. They also support residential rooftop solar by buying excess rooftop generation at a fair price.

California's Marin and Sonoma counties have already formed successful CCAs, which provide cleaner electricity at lower cost.[8]

With luck, the Los Angeles County CCA will join them soon. CCA is the way forward to 100% renewable electricity, and California is demonstrating this to the world.

Soccer

My family has a great time with good old youth soccer. Soccer gives my kids confidence, teaches them social skills, keeps them fit, binds them together as brothers, and brings them closer to their friends. They love it. All they need is a soccer ball and they'll play together blissfully for hours. And I love coaching, a great way for a parent to bond with a child.

Soccer unites and builds community. It brings joy and connection to people of all ages, all religions, and all ideologies, even in the poorest of places and the direst of circumstances.[9]

Thousands of others

There are thousands of other community networks within easy biking distance of my house. There are wonderful school groups, church groups, neighborhood groups, sports groups, city councils, new parent groups, death cafés, meetups, dances, house concerts, block parties, festivals...a whole biosphere of human community! The same is true where you live—but if you're interested in some network and a group doesn't exist in your area, consider starting it. Starting a group is a wonderful opportunity to serve others.

Race, privilege, and environmental equality

Interestingly, the community garden is racially diverse—whites are the minority there—but CCL membership is nearly 100% white, as is the international Transition Towns movement. The climate movement *must* find ways to appeal to a wide base that cuts across racial and social demographics. When this critical mass is achieved, climate action will become inevitable.

White people in the US, as a group, suffer far less exposure to environmental degradation than other racial groups. For example, people of color breathe air that is on average 40% more polluted.[10]

Why, then, is the mainstream environmental movement so thoroughly white?

One reason, perhaps, is that most of the resources in the environmental movement are controlled by Big Green NGOs, which are run like corporations with white chief executives compensated like corporate leaders. Many of these NGOs and their insider leaders aren't interested in rocking the boat; they defanged themselves decades ago in favor of replicating dominant institutional structures, of aligning with corporate power instead of resisting.[11] (Corporations, for their part, say *thanks for the greenwash*.)

But there are deeper reasons: privilege and positionality. First, privilege. Black and Hispanic households in the US have on average only ⅒ the wealth of whites—a staggering disparity.[12] To these households, the usual "solutions" pushed by climate activists are out of reach: solar panels, solar hot water heaters, electric cars, shopping at farmers' markets.

Just having the mental space to think about climate is a form of privilege, in the sense that doing so requires ample physical and psychological security, a degree of freedom from more immediate concerns like making rent. To many working-class people, including whites, thinking about climate change is for rich white people who live relatively high up on the hierarchy of needs.

Even many of my suggestions in this book, while not requiring money, nonetheless depend on other forms of privilege. Meditation requires carving out time away from one's job (or job search) and dependents. Deep composting requires continuous access to a patch of ground. Vegetarianism requires access to a variety of high-quality fruits and vegetables. Pulling good food out of dumpsters can lead to confrontations with police, which might be riskier for some than it is for me. Perhaps the only two of my changes that *don't* require privilege are quitting planes and bicycling.

Second, positionality. Consider how the issue of inclusivity is usually framed: how can we (privileged whites) get non-whites to join *our* movement? Implicit in this question is that the movement is a white movement, operating from a white perspective. Why would people of color want to join such a movement? Why would working-class whites want to join?

I have no simple answers to these questions—I doubt there is a simple answer—but I intuit that progress here will be critical for progress on climate. We (and here again, I'm talking to my white brothers and sisters) need to reach out to people who aren't like us, spend time with them, listen to and serve them—and win their trust. Instead of asking people of color to join us, we need to join them. We need to boldly but humbly go into their communities, give talks there, grow gardens there, route marches there. We need to invest our time and infrastructure there, fight for bike lanes and community gardens to go in there first, and bring in jobs and clean energy with community choice aggregation there. We need to let them experience how climate action improves their lives. Once they're *in*, we all win.

Money and gift

My formal communities—my town, state, nation, and work-place—feel somewhat impersonal to me. To a large extent, we have replaced the messiness of local personal community with a pseudo-community that runs on money and legal contracts (which ultimately are backed by violence). I can understand why: we need protection from bad actors, and constant bartering would be inconvenient. Still, I can't help feeling we've paid a high price for our efficient, legalistic, and impersonal society—which is, quite legally, destroying the biosphere. Is there a middle path here, too?

If there is, I suspect we'll find it through gift: helping others, and delighting them, while genuinely expecting nothing in return. Paradoxically, I find that I get back far more when I give freely. To receive something freely—*gratis*—leads to true gratitude; to give something freely, to true grace. The biosphere runs on grace.[13]

While the money economy emphasizes separation, the gift economy emphasizes connection. Even if gift transactions represent only a tiny sliver of the community's economy, they could catalyze a crucial shift in our mindset.

Peacemaking

It takes peacemakers to build and maintain community. Before we can be effective peacemakers, we must have peace ourselves.

The ego is a barrier to community. Our egos make us feel we need to be right, and to put ourselves above others. Any time a disagreement grows into a fight, with feelings of jealousy, anger, or hatred, *the root cause is always ego*. Left unchecked, these fights can turn into feuds capable of tearing apart the community.

Unless the community consists entirely of experienced meditators—and even when meditating correctly it can take a lifetime to dissolve the ego—negativity will arise. We therefore need to *make* peace.

I've found that the best thing to do when I feel negativity toward someone (for example, when I feel slighted, or angry) is to generate loving-kindness toward the other (*metta*) and to try seeing from his or her point of view. This action is usually enough. Many times, the negativity was only in my mind. Seeing from the other's point of view dissolves the negativity.

If this isn't enough, the next step is to talk with the person. This takes courage. Again, before talking to the person, and while talking to the person, I try to feel *metta*. While talking, I try to see the other person's point of view. If anger or other negativity starts to come up, I try to follow my breath. This can help me pause before I respond. Reacting mindlessly usually makes the situation worse. People are good at sensing egos; if I respond from my ego, the problem will only get worse.

When I was younger, I used to have enemies. Nothing was more important to me than being right. Sometimes I would nurture enmity for years at a time. This made me a slave to my ego. A few years ago, after many years of meditating, one day I noticed that I had no enemies, and hadn't for some time. All people are my friends. I now see that holding on to the need to be right made me miserable. Letting go of being right is a relief.

Listening

Good listening is like a pressure-relief valve for community. It lets people know they belong. It also provides the foundation upon which communities solve their problems.

When I listen to someone properly, there's nothing in the world more important than that person. My mind doesn't wander. I don't think about what I should say next while they talk so that I can sound smart. If I really listen, I don't need to talk. And if I do talk, when it's time for me to talk, it's with *metta* for the person I'm talking to.

Do I listen like this all the time? Unfortunately, I don't. But when I do, the connection is wonderful. This is a difficult practice, but the rewards are great. I find it helpful to maintain awareness of my body during conversation through breath or sensation.

Service

The opportunity to help someone is a precious thing. When you have this opportunity, jump on it. Conversely, be open to asking for help. If you ask for help, your needs will be met—and you'll give someone else an opportunity to help.

Leading and supporting

I used to need to be the leader. Instead of looking to contribute to the community and make it successful, I was just making my ego bigger.

Now, I look for supporting roles instead, and these feel good to me. I can see how much other members of the community have to offer. I can see their good qualities, and appreciate them, and feel sympathetic joy—joy in the success of others. This leads to a sense of harmony and connection.

When it is time for a leadership role, the leadership role will come. In my experience, at least, it is counterproductive to push for it. When it does come, it must be accepted with humility. The best leaders seek to serve. When the community is happy, the leader's reward is sympathetic joy.

Protest

I haven't seen much in the way of effective protest over the course of my life. Marches come and marches go; they're fun and they bring us together, remind us that we're not alone, but nothing changes.

When you study the unimaginably vast, cold void of space, it becomes harder to take the Earth for granted. Every patch of ground on this life-giving planet is sacred. Our native American brothers and sisters came from cultures that knew this, and many of them still remember. From their communities, a new template for protest is emerging: nonviolence rooted in the love of a place, nonviolence with a solid spiritual foundation. Protest needs a place. Protest needs community. It doesn't happen on social media. Protest is only as strong as the community that it builds—the soil out of which change can grow.

> If you want to be a rebel, be kind. Disobey with great love.
> — PANCHO RAMOS-STIERLE

Children

How can we talk to our children about our predicament? I've never sat down with my boys and said, "So, kids, time for you to know about global warming." Instead, I try to live a life that's informed by my knowledge and acceptance of global warming, a life that's consistent with my values. If my boys ask me something, I answer as honestly as I can. I certainly never go out of my way to scare them, but I don't lie to them, either.

Celebration

In industrial society, have we forgotten how to celebrate? Our modern holidays focus on consuming—sensual pleasure. But real celebration isn't about sensual pleasure. It's about losing one's

We're all in the same boat,
with death hanging over us,
doing our best to make sense
of this strange life.

sense of self and gaining a sense of something larger. This sense of something larger defies words, so we may as well call it *community*.

When I get this feeling of community, it often seems to involve people playing music and dancing. To be absorbed in this music and dance, to see everyone smiling and also absorbed, to me this is celebration. And for the music to be live is important; so much of our art and music today has been colonized, commodified. What thrills me is culture that comes as gift, as relationship.

Celebration is sacred, and the sacred is critically important. Celebration connects me somehow to humans long since dead, and humans who will be born far in the future. This makes death seem less terrifying. As for the people celebrating with me, in the present moment, I realize we're all in the same boat, with death hanging over us, doing our best to make sense of this strange life. In this way, celebration creates compassion.

Storytelling

It's not enough to change our own lives. It's not even enough to engage with the community. We must also be storytellers. Stories bind communities. They teach us, inspire us, and give us a way of making meaning out of the chaos of life. They direct our actions in powerful ways. Stories capture our imaginations, and global warming is the result of the greatest failure of imagination the world has ever seen.

Every one of us can tell this new story of living aligned with the biosphere, each other, and ourselves. This is the story of leaving fossil fuels far behind. This is the story of connection, of seeing ourselves within the biosphere and not above it.

How you tell it is up to you, whether through speech, song, poetry, or comedy; whether through teaching children, or running for office. But I can promise that no matter how you choose to tell the story, your first step will be to live it.

CHAPTER 16

Love

The real damage is done by those millions
who want to "survive...." Those people who
roll up their spirits into tiny little balls
so as to be safe. Safe? From what?
Life is always on the edge of death.
I choose my own way to burn.

— SOPHIE SCHOLL, satyagrahi (1921–1943)

Being the change is about transition. About going from being asleep to being awake. It's a story that has one foot in industrial civilization and one foot in whatever is coming next.

Writing this book has been like taking a snapshot of where I'm at on my path. Others have gone further: they are demonstrating that it's possible to live without any money or fossil fuels at all, and to be happier as a result. On my own path, as I continue to reduce, I'm actually experiencing increasing abundance. It's a good path.

As I keep walking on it, I get further from the core of industrial civilization. I'm heading toward its fringes. When I look back over my shoulder, what I'm walking away from looks pretty bad. It looks like we may have lost our way as a species. More and more, as I continue to walk, I'm able to see industrial civilization for what it was: an experiment or a phase that we had to go through, but that turned out not to work.

But now the results are in. Humans: let's stop burning fossil fuels. Let's stop killing each other and our planet. Let's stop merely talking about love; let's start *practicing* it. We have nothing to lose but our misery.

In order to embrace what's coming next, I had to let go of what went before. My grief was like the leap of a trapeze artist, letting go of one trapeze, flying through space, and catching the next one. There were times when tears poured down. I mourned the world I'd known my whole life. I mourned my children's future. I mourned how avoidable this all was. I mourned the strange and hard reality, and I mourned waking up. I mourned every blow struck in anger, and I mourned every bullet fired. I mourned all the species which are leaving us, never to return. I mourned this whole beautiful Earth.

But then, through these tears, I accepted reality as it is. Somehow, on the far side of the tears, I found the strength to go forward. Letting go has given me the space to imagine something new and better, a sea change. If you're grieving, let the tears flow. And when the tears finally stop, look around. You will see then that there are miracles all around you. And you will be inspired to work harder than you ever thought possible.

Let's work to build a world where everyone puts others above self and where we live aligned with the biosphere. In such a world, there is no war, crime, hatred, or negativity. Fundamentally, what's impossible about this vision? What law of physics does it break? I need to spend my time doing *something* while I'm here on this planet, and it might as well be to work toward this. And the place to start is with me.

I hope you'll join me. We can build this new world together. We can begin by observing our own selves by way of everyday physical sensations. We can begin by riding a bicycle.

I don't know how it's all going to turn out, but I have hope. The world has never seen a large-scale satyagraha, where instead of a handful of satyagrahi, there are millions. As we wake ourselves up, we also wake each other up. It's a mistake to assume we can't do this. We can—if we choose to.

Love and connection

Our ecological predicament challenges us for many reasons, not least of which is that it calls us to examine how we live. It calls us to recognize that we are part of the miraculous biosphere, not outside of it or above it; to accept, deeply, that we will die, and that death is also part of this miracle. Our bodies arise from this spinning, burning biosphere, and we mix back into it when we die. There is nothing to hold on to, and nothing to be afraid of.

In my darkest moments, when faced with the uncompromising reality of global warming in all of its surreal truth, I come back to my body. I feel my breath going in and out, or I observe a sensation, maybe some mild pain somewhere. The present moment contains the realization of how short my sojourn in this body really is. Each passing moment contains my death—a cosmic relay race of passing moments. And I realize that living aligned with my principles and doing everything I can for a better world is more important than staying safe.

The deepest spiritual revelations are all trivial at an intellectual level. For example: I'm made of matter, and so are you. We are both made of protons, neutrons, and electrons recycled from stellar explosions. Those building blocks spontaneously organized into ever-more-complex molecules, rising and passing away, transmuting from one form into another. Over hundreds of millions of years, the basic principles of encoding information into matter emerged. The biosphere began to produce a stunning stream of forms, billions upon billions of species—a process I feel certain is unfolding on uncountably many other worlds. Here on Earth, this awesomely beautiful stream of beings eventually included the first humans, and then my ancestors, and then my parents, and then me.

Yet what is this "me?" When I sit down and examine it carefully, dispassionately, it evaporates. I observe the matter in my body, and it is nothing but a flow, vibrating atoms, in a constant flux. I observe my mind and its thoughts, and they too are constantly

The universe is your partner.
The lizard in the woodpile is your partner.

changing. I observe the physical sensations that are the intersection of this mind and this body, and they also arise and pass away. I see how I crave pleasant sensations, and I see how the craving is misery. In the bright light of my calm observation, wanting dries up and blows away. I am not in control. I did not create myself. I am a calm observer with a front row seat at the greatest show imaginable: the universe itself.

How can I not smile as I watch this show? How can I be afraid when I realize how I formed? How can I feel separate from any other being when I am nature, when I am the biosphere, when I am the universe? How can I intentionally harm any being, once I've experienced the truth that all is connected? How can I put myself before others, when there is no "myself?"

To realize mindfulness in every moment is to experience that everything is sacred: the air we breathe; the food we eat. The stars and the oceans. Birds that fly and bacteria that decompose. A newly sprouted plant; a steaming pile of manure. The land we walk on. My body. Your body. This moment.

Learning to live respectfully within the biosphere is a sacred task. Learning to get along with each other is a sacred task. And learning how to be happy in our own minds, to be joyful on this Earth in the short time we're here, is a sacred task. These three sacred tasks are beautifully interconnected.

I wish that I could share with you the peace and the happiness that comes from having this concrete experience of connection, of non-self. But because it's a direct personal experience, I can't share it. I can only point to it clumsily with words.

Love and live lightly

Here's an idea that's simple and beautiful but goes against both the myths of the mainstream culture and our deepest mental habits. It's this: don't be afraid, and spread love every chance you get.

Don't be afraid—not even of death. Why are we afraid of death? Have we carefully examined this fear? Through meditation, I experience the simple truth that everything is constantly changing, including my own body and mind. It doesn't make sense to be attached to something so clearly in flux. As this grand fear dissolves, so do all the smaller ones. It also becomes very clear that fear is what stands in the way of spreading love.

Spread love every chance you get. Every time you encounter hate, stand up to it. This can take tremendous courage: humans have this terrible habit of killing those who spread love. Our culture is so full of fear that spreading love seems alien to most of us, alien to our story of hoarding, of not-enough.

This story can change. Indeed, I think we're already starting to change it. The key is for you and me to decide to live this way, and then to practice living this way. It takes work! But when others see us living according to a new story, that story will spread. The more I live like this, the happier I am; so I'll keep doing it in any case, whether or not it catches on.

Whatever you choose to do, do it in the spirit of dance: lightly, gracefully, with a smile, knowing well that this song will soon end, and a new song will start. The universe is your partner. The lizard in the woodpile is your partner. Enjoy it, and realize through your own experience that it's constantly changing. Everything is in a flux, in a flow, all matter and all thought. A single quark, your aging body, superclusters of galaxies: all is in a flow, a dance. It's all so beautiful, so full of love.

May all beings be happy.

Notes

Chapter 1: Waking Up

1. I will use the term *global warming* to refer to the increase in the average global temperature of recent decades and to the many changes to the Earth system caused by this warming, such as changes in precipitation. Strictly speaking, *climate change* is a more general phenomenon, of which today's global warming is but one example. The Earth has experienced many changes in climate over its 4.5 billions years of existence.

2. You can get this sense of plants as beings by standing silently by a beautiful tree, or by standing in front of a Van Gogh painting of trees, or by other means.

3. If you're reading this after 2023, the human population has likely surpassed eight billion.

4. I know that not many people see this yet, and I know I'm therefore taking a risk in saying it so clearly. But I know it to be true.

5. Mohandas Gandhi. *The Collected Works of Mahatma Gandhi*, Vol. 13 (1913), Ch. 153, page 241. [online]. gandhiserve.org/cwmg/VOL013 .PDF. Emphasis mine.

Chapter 2: Beyond *Green*

1. See, for example, the film *An Inconvenient Truth* directed by Davis Guggenheim and featuring Al Gore, 2006, DVD; or Bill McKibben. *Eaarth: Making a Life on a Tough New Planet*. Times Books, 2010.

2. Roving marauders would be attracted by hoarded food and ammunition. A better strategy for security might be to help build a strong community and to develop skills and relationships that make you indispensable.

3. Juliet Elperin. "White House solar panels being installed this week." *Washington Post*, August 15, 2013. [online]. washingtonpost.com /news/post-politics/wp/2013/08/15/white-house-solar-panels-finally -being-installed/.

Chapter 3: Global Warming: The Science

1. You can download the AR5: Intergovernmental Panel on Climate Change. "Assessment Reports." [online]. ipcc.ch/publications_and _data/publications_and_data_reports.shtml#1. The panel is divided into three working groups. Working Group I (WG1) presents the physical evidence for global warming and resulting changes occurring in the Earth system. Working Group II (WG2) presents the current and future impacts and human adaptation strategies. Working Group III (WG3) presents our understanding of the scientific, technological, environmental, economic, and social aspects of climate change mitigation and quantifies mitigation pathways, e.g., how much warming will occur under various paths available to humanity. WG1 assesses and summarizes the scientific literature, whereas WG2 and WG3 summarize the scientific and socioeconomic literature. Each working group provides a 30-page Summary for Policymakers (SPM). Subsequent citations use these abbreviations to identify sections within this report.

2. Here are some suggestions for further reading in order of increasing level of reader commitment. (1) Yoram Bauman and Grady Klein. *The Cartoon Introduction to Climate Change.* Island Press, 2014. (2) The 36-page overview US National Academy of Sciences and Royal Society. *Climate Change: Evidence and Causes.* National Academies Press, 2014. [online]. ap.edu/catalog/18730/climate-change -evidence-and-causes. (3) David Archer. *Global Warming: Understanding the Forecast,* 2nd ed. Wiley, 2011 (a college text for non-science majors). (4) "The Princeton Primers in Climate," a series of definitive but accessible books focusing on subtopics of climate, each by an expert in the subtopic: Princeton University Press. Catalogue Primers in Climate. [online]. press.princeton.edu/catalogs/series /princeton-primers-in-climate.html. I'd advise against trying to learn about climate science solely from the internet, as you'll have to wade through a great deal of misinformation and disjointedness. That said, you can find accurate (if disjointed) information at skepticalscience .com and realclimate.com.

3. Many scientists and humanists have given this new epoch a stratigraphic name, the *Anthropocene.* I hesitate to follow them for three reasons. First, the word has been embraced by "ecomodernists" and others who believe, with a blind faith, that technology is the way out of our predicament. (I disagree.) Second, it makes our destructive presence feel like a geologic fact, thereby potentially reducing political will to do what we can to reduce human impact on the biosphere.

Humanity still very much gets to choose just how bad global warming will get. Third, and perhaps most importantly, it presupposes that humans are the problem. In my opinion, *humans* aren't the problem. A particular human *culture* is the problem.

4. I surveyed my Earth scientist colleagues (participating in the survey required authorship on a peer-reviewed journal paper in Earth science) and received 66 responses. When asked "How often do you feel grief about global warming," only 10% responded "never" (a one on a scale from one to five) while half of respondents feel significant grief (with 14% and 35% choosing five and four on the scale, respectively).

5. IPCC AR5 WG1 SPM puts this temperature increase at 0.85°C, but this has since been revised upward by 24%: Mark Richardson et al. "Reconciled climate response estimates from climate models and the energy budget of Earth." *Nature Climate Change* 6 (2016). [online]. doi:10.1038/nclimate3066. Temperature will increase to about 1.2°C by 2020 (2015 was already 1.1°C to 1.3°C above the pre-industrial baseline, but this was just a single year), and 1.4°C by 2030. These projections are linear extrapolations based on the observed warming of 0.12°C per decade between 1951 and 2012.

6. Camilo Mora et al. "The projected timing of climate departure from recent variability." *Nature* 502 (2013). doi:10.1038/nature12540.

7. Models are tools for predicting how variables might change in the future. A global climate model is software code that runs on supercomputers and represents physical, chemical, and biological processes in the atmosphere, ocean, land, and ice. Earth system models are global climate models that explicitly model the carbon cycle. These models represent the Earth by dividing it into three-dimensional grid cells; horizontal resolution is typically about 100 km, although this is falling as computers get faster. At each time step, variables in every grid cell (such as temperatures, cloud amounts, sea ice amounts, etc.) are updated based on values from the previous time step and values from neighboring grid cells.

8. The business-as-usual scenario, named "RCP 8.5," was described in Keywan Riahi et al. "RCP 8.5: A scenario of comparatively high greenhouse gas emissions." *Climatic Change* 109(1–2) (2011). [online]. doi:10.1007/s10584-011-0149-y. The mitigation scenario, named "RCP 4.5," is described in Allison M. Thomson et al. "RCP 4.5: A pathway for stabilization of radiative forcing by 2100." *Climatic Change* 109(1–2) (2011). [online]. doi:10.1007/s10584-011-0151-4.

9. These results used a model experiment called "historicalNat" to produce estimates of the range of background variability. In the

historicalNat experiment, the models ran from 1860 to 2005 with no anthropogenic CO_2 emissions, allowing for a pure comparison with the two RCP runs. However, only 17 climate models actually ran historicalNat, whereas 39 models ran an experiment called "historical" which included observed changes in atmospheric composition, including anthropogenic CO_2 emissions. Mora et al. also report results using the historical experiment as the background variability. Because this already includes some warming, the climate departure dates relative to it are delayed relative to the historicalNat results, to 2047 for RCP 8.5 (this is a mean of the 39 models, with a standard error of three years) and 2069 for RCP 4.5 (with a standard error of four years). However, these results are clearly biased due to the anthropogenic warming in the background. The results from the historicalNat do not have this bias. Mora et al. chose to highlight the biased results in order to be more conservative; in my opinion, this was a mistake. In science, it's always best to report whatever is closest to the truth, to the best of your knowledge.

10. Note that surface temperature is just one variable in the Earth system. Any variable can be analyzed for anthropogenic departure. For example, global departure has already occurred for ocean surface acidity from anthropogenic CO_2 dissolved in the ocean.

11. I therefore suggest we mitigate as if our lives depend on it. I personally don't think there's anything more important for humanity to do at this time.

12. Thanks to Jan Sedlacek for providing the underlying multi-model mean data, from which I created this black-and-white version of Figure 12.5 from IPCC AR5 WG1 (a.k.a. Matthew Collins et al. "Long-term Climate Change: Projections, Commitments and Irreversibility" in T. F. Stocker et al., eds. *Climate Change 2013: The Physical Science Basis. Working Group I Contribution to the Fifth Assessment Report of the Intergovernmental Panel on Climate Change*. Cambridge University Press, 2013, p. 1039).

13. The number gives the approximate radiative forcing in watts per square meter (W/m^2) in the year 2100 for the scenario. We'll discuss radiative forcing in detail later in this chapter.

14. Jasper van Vliet et al. "Meeting radiative forcing targets under delayed participation." *Energy Economics* 31 (2009). [online]. doi:10.1016/j.eneco.2009.06.010.

15. See IPCC AR5 WG3 Chapter 12, Table 12.2. Predictions are the multi-model means. Uncertainties are one standard deviation of the multi-model distribution. Note that emissions predictions beyond 2100 require extended RCP scenarios, which make simple (and possibly

simplistic) assumptions about greenhouse gas and aerosol emissions beyond 2100. As model predictions extend further into the future, they naturally become increasingly uncertain.

16. J. D. Annan and J. C. Hargreaves, J. C. "A new global reconstruction of temperature changes at the Last Glacial Maximum." *Climate of the Past* 9 (2013). [online]. doi: 10.5194/cp-9-367-2013.

17. Lorraine E. Lisiecki and Maureen E. Raymo. "A Pliocene-Pleistocene stack of 57 globally distributed benthic $\delta^{18}O$ records." *Paleoceanology* 20(1) (2005). [online]. doi:10.1029/2004PA001071.

18. James Hansen et al. "Climate sensitivity, sea level and atmospheric carbon dioxide." *Philosophical Transactions of the Royal Society A* 371 (2013). [online]. doi:10.1098/rsta.2012.0294.

19. A key reason the swaths in Figure 3.2 are so wide is due to a "known unknown," our uncertainty about how clouds work. (One of my research interests is low-altitude clouds: how they interact with the Earth system, how they change as the planet warms, and how their changes in turn affect warming.) The global models used to make climate projections divide the Earth into grid cells that are currently about a degree latitude and a degree longitude in size—much larger than individual clouds. This means the models must statistically approximate cloud variables, such as the total cloud cover at different altitudes, in each grid cell. Each model does this differently, and one result of this is that the models differ on how clouds interact with atmospheric dynamics, and how they'll change as the planet continues to warm. For example, some models predict an increase in low clouds with warming, while others predict a decrease. Low clouds (such as the stratocumulus clouds on an overcast day) cool the planet by reflecting sunlight back to space: models with more low clouds tend to predict cooler temperatures, and models with fewer low clouds predict warmer temperatures.

20. This temperature would be set by a simple balance between absorbed solar radiation and emitted thermal infrared radiation. There would not even be clouds reflecting sunlight and interacting with outgoing infrared light, because water vapor is among the greenhouse gases we've just imagined away. The Earth's actual average surface temperature, with the greenhouse effect, is about 15°C.

21. Inferred from data from NOAA Earth System Research Laboratory. "Trends in Atmospheric Carbon Dioxide." [online]. esrl.noaa.gov /gmd/ccgg/trends/global.html.

22. The rate of heat conduction between two sides of a conducting object (like our blanket) is proportional to the difference in temperature between them.

23. This absorption is quantum-mechanical. Infrared photons coming up from the Earth have a range of frequencies following the Planck spectrum. When an infrared photon hits a molecule of water or CO_2 (or another greenhouse gas) with the quantum of energy needed to excite the molecule from its ground state to e.g., a bending vibrational state (i.e., molecules can only absorb photons of certain frequencies, but the photon happens to have one of these quantum-mechanically allowed frequencies), it can be absorbed by the molecule, which then starts to vibrate. After some time, the molecule will de-excite and emit a photon in a random direction, possibly out to space. However, in the lower atmosphere, the molecule is more likely to collide with some other molecule first. When this collision occurs, it can transfer energy to the molecule it collides with. The net result is that most of the upwelling infrared photons from the Earth's surface fail to escape into space, and instead warm up the lower atmosphere.

24. Measuring the global mean reflected solar energy from space is challenging; satellites records show no clear trend. While you might think melting snow and ice would cause more solar energy to be absorbed (and this does happen regionally), cloud patterns can change and compensate in the global mean.

25. The residence time is an estimate of how long it would take for the excess over preindustrial levels to be halved. This doesn't mean the next halving will take the same amount of time (i.e., the decay is not necessarily a single exponential process). CO_2 is chemically inert and has a long residence time, while CH_4 is chemically reactive and has a short residence time.

26. These values are from Drew T. Shindell et al. "Improved attribution of climate forcing to emissions." *Science* 326 (2009), at p. 716. [online]. doi:10.1126/science.1174760. For a given gas, GWP estimates depend on which direct and indirect warming effects from that gas are included in the estimate. For example, the IPCC GWP estimates for methane do not include gas-aerosol interactions, in which methane suppresses the formation of aerosols which cool the climate; excluding this effect, the IPCC AR5 estimates GWP_{20} of only 86.

27. PCC AR5 WG1 Chapter 8, p. 714 gives a GWP_{20} of 264 and a GWP_{100} of 265 with uncertainties 20% and 30% respectively.

28. Estimates of percentage of current greenhouse radiative forcing in this section are made by dividing the forcing in question by the total positive forcing from greenhouse gases: $CO_2 + CH_4 +$ halocarbons $+ N_2O + CO +$ NMVOC $= 3.33$ W/m^2. Note that I've omitted NMVOC from the discussion for simplicity (they contribute 0.1 [0.05 to 0.15] W/m^2 and are declining); for details see IPCC AR5 WG1 Chapter 2.

29. IPCC AR5 WG1 Chapter 8, p. 731.
30. This 5% factors in the cooling effect of ozone destruction.
31. Shindell et al. "Improved attribution of climate forcing to emissions."
32. This includes formation of ozone and stratospheric water vapor, but not gas-aerosol interactions.
33. IPCC AR5 WG3 SPM.
34. From my own calculation using a methane GWP of 105, based on the 16% figure for a methane GWP of 34, with the other greenhouse gas GWPs fixed.
35. IPCC AR5 WG1 Chapter 2, p. 167.
36. IPCC AR5 WG1 Chapter 6, p. 541.
37. Note that the IPCC estimate of methane emissions from fossil fuel production relies on a 2012 EPA estimate: US EPA, Office of Atmospheric Programs. "Global anthropogenic non-CO_2 greenhouse gas emissions: 1990–2030." EPA Report # EPA 430-R-12-006. [online]. epa.gov/climatechange/Downloads/EPAactivities/EPA_Global _NonCO2_Projections_Dec2012.pdf. Newer studies are finding that this estimate may be significantly too low.
38. The 2002 to 2011 annual average.
39. IPCC AR5 WG1 SPM. Between 2002 and 2011, on average humans emitted 8.3 [7.6–9.0] gigatonnes of carbon (GtC) per year from burning fossil fuels and making cement, and 0.9 [0.1–1.7] GtC per year from land-use change. Cement production accounts for about 4% of human CO_2 emissions: IPCC AR5 WG1 Chapter 6, p. 489.
40. IPCC AR5 WG1 Chapter 6, p. 486.
41. The measurement process is described on the website of the Carbon Cycle Greenhouse Gases Group of the Global Monitoring Division of the Earth System Research Laboratory of the National Oceanic and Atmospheric Administration (online at esrl.noaa.gov/gmd/ccgg/). In a nutshell: Scientists at the top of the mountain measure the CO_2 fraction in dried air via infrared absorption. Infrared light shines into a glass tube containing air. An infrared detector on the far side of the tube measures the transmitted infrared light. CO_2 blocks infrared light (which is also why it warms the Earth), so the more CO_2 in the air, the less infrared light makes it to the detector. The output of the detector is a voltage which increases with the power of the incident infrared radiation. Once calibrated, that voltage can be converted to the CO_2 fraction to within 0.2 ppmv (parts per million by volume).

    The trickiest part of the measurement is calibration: accurately and precisely converting the detector voltage to the fraction of CO_2 in the air. This is done with three reference mixtures of air, themselves carefully calibrated, which are turned on once per hour for four minutes each. The three data points are fitted quadratically, giving the conversion function. Systematic errors are guarded against by checking prepared "target" samples of air of various and known CO_2 fractions, and by sending flasks of air to National Institute of Standards and Technology (NIST) in Boulder, Colorado, for independent measurement.

42. Data are from US NOAA. "A Global Network for Measurements of Greenhouse Gases in the Atmosphere." [online]. esrl.noaa.gov/gmd /ccgg/.

43. I used the function $y = 280 + (1 + a)^{t-b}$. The best fit values (on annual mean values from 1959 to 2016) are b = 1790 and a = 0.0217.

44. The record going back to about 400,000 years ago comes from an ice core taken near Vostok Station, Antarctica: J. R. Petit et al. "Climate and atmospheric history of the past 420,000 years from the Vostok Ice Core, Antarctica." *Nature* 399 (1999). [online]. doi:10.1038/20859. The CO_2 fraction was measured by gas chromatography. Because this is a "proxy" record, it has error bars in *both* axes (time and CO_2 fraction). The error bar in the absolute time (the x-axis) is less than ±15 ky (kiloyears) over the whole record, and less than ±5 ky over the last 110,000 years. The error bar in the CO_2 fraction (the y-axis) is ±3 ppmv. The record going back to about 800,000 years ago comes from an ice core taken on Dome C, Antarctica: Lüthi et al. "High-resolution carbon dioxide concentration record 650,000–800,000 years before present." *Nature* 453 (2008). [online]. doi:10.1038 /nature06949. The record from 1,000 years ago until almost the present day comes from an Antarctic ice core taken on the Law Dome: Etheridge et al. "Natural and anthropogenic changes in atmospheric CO_2 over the last 1000 years from air in Antarctic ice and firn." *Journal of Geophysical Research* 101 (1996). [online]. doi:10.1029/95JD03410.

45. Although the error in absolute time at this point in the record is around ±5 ky, the error in time *durations* is much smaller.

46. Data from IPCC AR5 WG3, Chapters 7–11. Note that this table preserves the GWP_{100} basis used by the IPCC. In other words, methane is assigned the relatively low GWP of 21, arguably making it underrepresented in this table.

47. M. MacLeod et al. *Greenhouse gas emissions from pig and chicken supply chains—A global life cycle assessment.* Food and Agriculture Organization of the United Nations (FAO), 2013. [online]. fao.org /docrep/018/i3460e/i3460e.pdf; C. Opio et al. *Greenhouse gas emissions from ruminant supply chains: A global life cycle assessment.* Food and Agriculture Organization of the United Nations (FAO), 2013. [online]. fao.org/docrep/018/i3461e/i3461e.pdf.

48. IPCC AR5 WG3 Chapter 8, p. 605.

49. The data are from the Berkeley Earth. Land + Ocean surface temperature time series: Berkley Earth. *Land + Ocean Data.* [online]. berkeleyearth.org/land-and-ocean-data/.

50. Robert Rohde et al. "Berkeley Earth temperature averaging process." *Geoinformatics & Geostatistics: An Overview* 1:2 (2013). [online]. scitechnol.com/berkeley-earth-temperature-averaging-process -IpUG.pdf. The ocean data are from HadSST: Asia-Pacific Data-Research Centre. *Data documentation: Hadley Centre SST data set (HadSST).* [online]. apdrc.soest.hawaii.edu/datadoc/hadsst.php.

51. Mark Richardson et al. "Reconciled climate response estimates from climate models and the energy budget of Earth." *Nature Climate Change* 6 (2016). [online]. doi:10.1038/nclimate3066.

52. Other data sets tell the same story. According to the NASA data set, for example, 16 of the 17 warmest years on record occurred between 2001 and 2016: US NASA. "GISS Surface Temperature Analysis (GISTEMP)." [online]. data.giss.nasa.gov/gistemp/. For a summary article: Justin Gillis. "Earth sets a temperature record for the third straight year." *New York Times,* January 18, 2017. [online]. nytimes.com/2017/01/18/science/earth-highest-temperature -record.html.

53. Perhaps if you're reading this in 2025, you're thinking back, wistfully, to that much cooler year, 2016. When I read something about climate change written in the past, I often find myself thinking this.

54. Sydney Levitus et al. "Anthropogenic warming of Earth's climate system." *Science* 292(5515) (2001). [online]. doi:10.1126/science.1058154.

55. Data from NOAA, Ocean Climate Laboratory, Global Ocean Heat and Salt Content. "Basin time series of heat content (product, 0–2000 meters)." [online]. nodc.noaa.gov/OC5/3M_HEAT _CONTENT/basin_data.html. Data described in S. Levitus et al. "World ocean heat content and thermosteric sea level change (0–2000 m), 1955–2010." *Geophysical Research Letters* 39 (2012). [online]. doi:10.1029/2012GL051106.

56. Noah Diffenbaugh and Christopher Field. "Changes in ecologically critical terrestrial climate conditions." *Science* 341(6145) (2013). doi:10.1126/science.1237123.

57. IPCC AR5 WG1 SPM. Over these same periods, the rate of Antarctic ice sheet loss has increased from 30 Gt per year to 147 Gt per year, a four-fold increase.

58. Fiammetta Straneo and Patrick Heimbach. "North Atlantic warming and the retreat of Greenland's outlet glaciers." *Nature* 504 (2013). [online]. doi:10.1038/nature12854.

59. IPCC AR5 WG1 SPM. Some of this sea level rise is from ice melt, and some is from thermal expansion of water.

60. Christopher S. Watson et al. "Unabated global mean sea-level rise over the satellite altimeter era." *Nature Climate Change* 5 (2015). [online]. doi:10.1038/nclimate2635.

61. IPCC AR5 WG1 SPM.

62. Ibid.

63. Additional changes include considerable reduction in Siberian permafrost thickness and extent; decreasing northern hemisphere snow covered area (June snow cover is decreasing by 12% per decade); and non-surface atmospheric warming in the lower atmosphere as measured by satellites. I have to be honest: to me, these and other changes seem surreal, like bad science fiction. But they're as real and as verifiable as a melting ice cube.

64. I made this version of figure SPM.5 from IPCC AR5 WG1 SPM. Source: T. F. Stocker et al., eds. *Climate Change 2013*.

65. Mike Lockwood. "Solar Influence on Global and Regional Climates." *Surveys in Geophysics* 33(3) (2012). [online]. doi:10.1007/s10712-012 -9181-3.

66. Note that this cooling from ozone destruction by halocarbons represents a direct connection between the Antarctic ozone hole and global warming. The magnitude of the negative forcing from global depletion of the ozone layer is about 5% of the magnitude of the net forcing (which, of course, is positive), and some of this negative forcing is from the ozone hole. In this sense the ozone hole does play a role in global warming, albeit a small one.

67. David Herring. "Earth's Temperature Tracker." NASA. Earth Observatory website, November 5, 2007. [online]. earthobservatory.nasa .gov/Features/GISSTemperature/giss_temperature.php/.

68. But what *is* black carbon? Strangely, no lab has a sample of black carbon in a vial, and there is no agreed-upon definition for the substance, which can perhaps best be described as "light-absorbing refractory carbonaceous matter of uncertain character." For example: P. R. Buseck et al. "Are black carbon and soot the same?" *Atmospheric Chemistry and Physics Discussions* 12 (2012). [online]. doi:10.5194/acpd -12-24821-201.

69. Kristina Pistone et al. "Observational determination of albedo decrease caused by vanishing Arctic sea ice." *Proceedings of the National Academy of Sciences* 111(9) (2014). [online]. doi:10.1073/pnas .1318201111.

70. IPCC AR5 WG1 Chapter 7, p. 592. In 2015, the cloud response was

still the largest source of uncertainty in estimating climate sensitivity. Three positive feedbacks are known with varying degrees of confidence. The most robust is an increase in high cloud top height with warming. This is a positive feedback because higher clouds trap more infrared radiation. The second positive cloud feedback, accepted with medium confidence (IPCC AR5 WG1 Chapter 7, p. 589) is a global shift of cloud patterns to the poles, where there's less sunlight; this decreases albedo. The third positive cloud feedback, accepted with low confidence, is a decrease in subtropical low clouds. Global models give a wide range of magnitudes, and a few even give a negative feedback.

71. Graeme L. Stephens et al. "The albedo of Earth." *Reviews of Geophysics* 53 (2015). [online]. doi:10.1002/2014RG000449.

72. Yadong Sun et al. "Lethally hot temperatures during the early Triassic greenhouse." *Science* 338(6105) (2012). [online]. doi:10.1126/science .1224126.

73. Reservoir size estimates in Table 3.3 are based on estimates from IPCC AR5 WG1 Chapter 6, and P. Falkowski et al. "The global carbon cycle: A test of our knowledge of Earth as a system." *Science* 290(5490) (2000). [online]. doi:10.1126/science.290.5490.291.

74. IPCC AR5 WG1 SPM, p. 12.

75. David Archer. *The Global Carbon Cycle*. Princeton, 2010. For quantitative details about the carbon cycle response to increased CO_2 and warming, see IPCC AR5 WG1 Chapter 6, Figure 6.20 and accompanying text.

76. IPCC AR5 WG1 Chapter 6, p. 492. In discussions of the carbon cycle, the conventional unit is gigatonnes of carbon (GtC, $1 \text{ Gt} = 10^9 \text{ t}$) or equivalently petagrams of carbon (PgC, $1 \text{ Pg} = 10^{15} \text{ g}$). Elsewhere in the book, I may measure the mass of CO_2 instead of the mass of the carbon atoms in CO_2. One GtC = 3.67 $GtCO_2$.

77. IPCC AR5 WG1 SPM, p. 12.

78. S. Khatiwala et al. "Reconstruction of the history of anthropogenic CO_2 concentrations in the ocean." *Nature* 462 (2009). [online]. doi:10.1038/nature08526.

79. Archer. *The Global Carbon Cycle*, p. 177.

80. W. Kolby Smith et al. "Large divergence of satellite and Earth system model estimates of global terrestrial CO_2 fertilization." *Nature Climate Change* 6 (2016). [online]. doi:10.1038/nclimate2879.

81. T. W. Crowther et al. "Quantifying global soil carbon losses in response to warming." *Nature* 540 (2016). [online]. doi:10.1038 /nature20150.

82. Craig D. Allen et al. "A global overview of drought and heat-induced tree mortality reveals emerging climate change risks for forests." *Forest Ecology and Management* 259 (2010). [online]. doi:10.1016/j.foreco.2009.09.001.

83. Paulo Montiero Brando et al. "Abrupt increases in Amazonian tree mortality due to drought-fire interactions." *Proceedings of the National Academy of Sciences* 111(17) (2014). [online]. doi:10.1073/pnas.1305499111.

84. Ibid. and Oliver L. Phillips et al. "Drought sensitivity of the Amazon rainforest." *Science* 323(5919) (2009). [online]. doi:10.1126/science.1164033.

85. Pierre Friedlingstein et al. "Uncertainties in CMIP5 Climate Projections due to Carbon Cycle Feedbacks." *Journal of Climate* 27 (2014). [online]. doi:10.1175/JCLI-D-12-00579.1.

86. David Archer et al. "Ocean methane hydrates as a slow tipping point in the global carbon cycle." *Proceedings of the National Academy of Sciences* 106(49) (2009). [online]. doi:10.1073/pnas.0800885105.

87. Archer. *The Global Carbon Cycle*, p. 178.

88. IPCC AR5 WG1 Chapter 6, p. 467. The IPCC cites 9.5±0.8 GtC, but I have subtracted cement production, which accounted for roughly 4% of total CO_2 emissions in 2000–2009 (IPCC AR5 WG1 Chapter 6, p. 489). The 2002–2011 average (including cement production) was 8.3 GtC per year.

89. IPCC AR5 WG3 Chapter 5, p. 357.

90. Data from IPCC AR5 WG1 Chapter 6, Table 6.1 and Figure 6.1.

91. The formula is $CO_2 + CO_3^{2-} + H_2O \leftrightarrow 2HCO_3^-$.

92. The formula is $CaCO_3 \leftrightarrow Ca^{2+} + CO_3^{2-}$. Because the CO_2 is reacting with carbonate ions, CO_3^{2-}, this reaction is pushed to the right.

93. I've taken data from the BP corporation's *Statistical Review of World Energy 2012*, which goes back to 1965. For the current edition, see BOP Global. *Statistical Review of World Energy*. [online]. bp.com/en/global/corporate/energy-economics/statistical-review-of-world-energy.html.

94. Increasing airborne CO_2 by 7.8 gigatonnes is equivalent to raising the atmospheric fraction by 1 ppm. (Note that emitting a certain amount of CO_2 is not the same as increasing airborne CO_2 by that amount, since some of the emissions would go into other carbon reservoirs, the land and ocean sinks.)

95. According to IPCC AR5 WG1 SPM, from 1750 to 2011, humans emitted 375 [345 to 405] GtC from fossil fuel burning and 180 [100 to 260] GtC from deforestation. A more careful analysis would incorporate

actual deforestation and cement production CO_2 emission data as a function of time.

96. IPCC AR5 WG1 Chapter 6, p. 493.

97. Svante Arrhenius presented his paper, "On the Influence of Carbonic Acid in the Air Upon the Temperature of the Ground," in 1895.

98. Rare heavy water molecules (containing either ^{18}O or D, heavier isotopes of ^{16}O and H respectively) evaporate at lower rates and condense at higher rates than H_2O at a given temperature, and these rate differences become more pronounced as temperature decreases. Knowledge of the temperature relationship of these rates therefore allows us to estimate the Earth's temperature at past times. See J. Jouzel et al. "Orbital and millennial Antarctic climate variability over the past 800,000 years." *Science* 317(5839) (2007). [online]. doi:10.1126/science.1141038.

99. Dome C data from: NOAA World Data Center for Paleoclimatology. "Ice Core." [online]. ncdc.noaa.gov/paleo/icecore/antarctica/domec/domec_epica_data.html.

100. The driver of climate change over this 800,000-year period was subtle periodic changes in the Earth's orbit.

101. Note that CO_2 changes of only about 80 ppm corresponded in the ice core records to Antarctic surface temperature changes of about 12°C. By comparison, as of 2016 we've already increased the atmospheric CO_2 concentration by 120 ppm over preindustrial levels, and the Earth system is still in the process of adjusting to a new equilibrium, with positive feedbacks kicking in and more warming in the pipeline. This might seem to imply a 12°C increase, but remember, the Earth system is complicated. The observed 12°C increases occurred when an ice-covered Earth came out of glacial periods, and about ⅔ of the warming was due to albedo change as ice melted. Today, we're not in a glacial period, so we won't experience such a large albedo amplification; our primary forcing is from greenhouse gases, which accounted for about ⅓ of the glacial/interglacial amplification. See RealClimate (Eric Steig). "The lag between temperature and CO_2." April 27, 2007. [online]. realclimate.org/index.php/archives/2007/04/the-lag-between-temp-and-co2/.

Chapter 4: Global Warming: The Outlook

1. For a more complete discussion of impacts, please see the global warming books I listed in Chapter 3, note 2, or the IPCC AR5 WG2 report.

2. IPCC AR5 WG1 SPM.

3. Jean-Marie Robine et al. "Death toll exceeded 70,000 in Europe during the summer of 2003." *Comptes Rendus Biologies* 331 (2) (2008). [online]. doi:10.1016/j.crvi.2007.12.001.

4. Peter A. Stott et al. "Human contribution to the European heatwave of 2003." *Nature* 432 (2004). [online]. doi:10.1038/nature03089.

5. Nikolaos Christidis et al. "Dramatically increasing chance of extremely hot summers since the 2003 European heatwave." *Nature Climate Change* 5 (2015). [online]. doi:10.1038/nclimate2468.

6. I-Ching Chen et al. "Rapid range shifts of species associated with high levels of climate warming." *Science* 333(6045) (2011). [online]. doi:10.1126/science.1206432; Kai Zhu et al. "Failure to migrate: Lack of tree range expansion in response to climate change." *Global Change Biology* 18(3) (2011). [online]. doi:10.1111/j.1365-2486.2011.02571.x; Harald Pauli et al. "Recent plant diversity changes on Europe's mountain summits." *Science* 336(6079) (2012). [online]. doi:10.1126/science.1219033.

7. Elvira S. Poloczanska et al. "Global imprint of climate change on marine life." *Nature Climate Change* 3 (2013). [online]. doi:10.1038/nclimate1958.

8. Chris D. Thomas et al. "Extinction risk from climate change." *Nature* 427 (2004). [online]. doi:10.1038/nature02121.

9. Francesca A. McInerney and Scott L. Wing. "The Paleocene-Eocene Thermal Maximum: A Perturbation of Carbon Cycle, Climate, and Biosphere with Implications for the Future." *Annual Review of Earth and Planetary Sciences* 39 (2011). [online]. doi:10.1146/annurev-earth-040610-133431.

10. E. O. Wilson. *Half-Earth: Our Planet's Fight for Life*. Liveright, 2016, p. 14.

11. Mark C. Urban. "Accelerating extinction risk from climate change." *Science* 348(6234) (2015). [online]. doi:10.1126/science.aaa4984.

12. Mora et al. "The projected timing of climate departure."

13. IPCC AR5 WG1 SPM.

14. Ibid.

15. Andreas F. Prein et al. "The future intensification of hourly precipitation extremes." *Nature Climate Change* 7 (2017). [online]. doi:10.1038/nclimate3168.

16. IPCC AR5 WG1 Chapter 2, p. 204.

17. IPCC AR5 WG1 Chapter 2, p. 223.

18. At the time of this writing, a definitive global warming attribution for specific droughts has not been made, but evidence is mounting.

19. IPCC AR5 WG1 Chapter 2, p. 227.

20. Benjamin I. Cook, Toby R. Ault, and Jason E. Smerdon. "Unprec-

edented 21st century drought risk in the American Southwest and Central Plains." *Science Advances* 1(1) (2015). [online]. doi:10.1126 /sciadv.1400082.

21. Daniel Griffin and Kevin J. Anchukaitis. "How unusual is the 2012–2014 California drought?" *Geophysical Research Letters* 41 (2014). [online]. doi:10.1002/2014GL062433.

22. Mike McPhate. "California today: More than 100 million trees are dead. What now?" *New York Times,* November 21, 2016. [online]. nytimes.com/2016/11/21/us/california-today-dead-trees-forests.html.

23. John T. Abatzoglou and A. Park Williams. "Impact of anthropogenic climate change on wildfire across western US forests." *Proceedings of the National Academy of Sciences* 113(42) (2016). [online]. doi:10.1073 /pnas.1607171113.

24. Tatiana Schlossberg. "Climate change blamed for half of increased fire danger." *New York Times,* October 10, 2016. [online]. nytimes.com /2016/10/11/science/climate-change-forest-fires.html.

25. IPCC AR5 WG2 SPM.

26. Corey Watts. "A brewing storm: The climate change risks to coffee." Climate Institute, 2016. [online]. climateinstitute.org.au/coffee.html. Climate-related threats to the global coffee crop include drought, rising temperatures, and increases in pests and disease.

27. IPCC AR5 WG2 Chapter 5, p. 364.

28. Ibid.

29. Stephane Hallegatte et al. "Future flood losses in major coastal cities." *Nature Climate Change* 3 (2013). [online]. doi:10.1038/nclimate1979.

30. Ian Urbina. "Perils of climate change could swamp coastal real estate." *New York Times,* November 24, 2016. [online]. nytimes .com/2016/11/24/science/global-warming-coastal-real-estate.html.

31. IPCC AR5 WG2 Chapter 5, p. 364.

32. Eric Rignot et al. "Widespread, rapid grounding line retreat of Pine Island, Thwaites, Smith, and Kohler glaciers, West Antarctica, from 1992 to 2011." *Geophysical Research Letters* 41(10) (2014). [online]. doi:10.1002/2014GL060140; Johannes Feldmann and Anders Levermann. "Collapse of the West Antarctic Ice Sheet after local destabilization of the Amundsen Basin." *Proceedings of the National Academy of Sciences* 112(46) (2015). [online]. doi:10.1073/ pnas.1512482112.

33. Feldmann and Levermann. "Collapse of the West Antarctic Ice Sheet after local destabilization of the Amundsen Basin." Ian Joughin et al. "Marine ice sheet collapse potentially under way for the Thwaites Glacier Basin, West Antarctica." *Science* 344(6184) (2014). [online]. doi:10.1126/science.1249055.

34. Robert M. DeConto and David Pollard. "Contribution of Antarctica to past and future sea-level rise." *Nature* 531 (2016). [online]. doi:10.1038/nature17145.

35. Colin P. Kelley et al. "Climate change in the Fertile Crescent and implications of the recent Syrian drought." *Publications of the National Academy of Sciences* 112(11) (2015). [online]. doi:10.1073/pnas .1421533112; John Wendle. "The ominous story of Syria's climate refugees." *Scientific American,* December 17, 2015. [online]. scientific american.com/article/ominous-story-of-syria-climate-refugees/.

36. Solomon M. Hsiang et al. "Quantifying the influence of climate on human conflict." *Science* 341(6151) (2013). [online]. doi:10.1126 /science.1235367.

37. W. J. Hennigan. "Climate change is real: Just ask the Pentagon." *Los Angeles Times,* November 11, 2016. [online]. latimes.com/nation/la-na -military-climate-change-20161103-story.html.

38. IPCC AR5 WG2 SPM.

39. IPCC AR5 WG1 Chapter 6, p. 544.

40. Archer. *The Global Carbon Cycle.*

41. McInerney and Wing. "The Paleocene-Eocene Thermal Maximum."

42. James W. Kirchner and Anne Weil. "Delayed biological recovery from extinctions throughout the fossil record." *Nature* 404 (2000). [online]. doi:10.1038/35004564.

43. In the survey of Earth scientists I mentioned in Chapter 3, when asked "Does it bother you, personally, to burn fossil fuels?" 52% said "yes," 41% said "sometimes," and 7% said "no" (58 responses).

44. According to the Royal Society: *Geoengineering the climate: Science, governance and uncertainty.* RS Policy document 10/09, 2009. [online]. royalsociety.org/topics-policy/publications/2009/geo engineering-climate/.

45. Lyla L. Taylor et al. "Enhanced weathering strategies for stabilizing climate and averting ocean acidification." *Nature Climate Change* 6 (2016). [online]. doi:10.1038/nclimate2882.

46. It takes about two tonnes of silicate rock to remove one tonne of CO_2. But one tonne of carbon mined from the ground releases about four tonnes of CO_2 gas when burned. So every tonne of carbon mined would require about eight tonnes of silicate rock to mitigate.

47. Ziahua Liu et al. "Atmospheric CO_2 sink: Silicate weathering or carbonate weathering?" *Applied Geochemistry* Vol. 26 (Supplement) (2011). [online]. doi: dx.doi.org/10.1016/j.apgeochem.2011.03.085.

48. Royal Society. *Geoengineering the climate,* p. 49.

49. Graeme L. Stephens et al. "The albedo of Earth." *Reviews of Geophysics* 53 (2015). [online]. doi:10.1002/2014RG000449.

50. H. Damon Matthews and Ken Caldeira. "Transient climate-carbon simulations of planetary geoengineering." *Publications of the National Academy of Sciences* 104(24) (2007). [online]. doi: 10.1073pnas.0700419104.

51. Royal Society. *Geoengineering the climate*, p. 32.

52. R. Lal. "Soil carbon sequestration impacts on global climate change and food security." *Science* 304(5677) (2004). [online]. doi:10.1126/science.1097396.

53. David S. Powlson et al. "Limited potential of no-till agriculture for climate change mitigation." *Nature Climate Change* 4 (2014). [online]. doi:10.1038/NCLIMATE2292.

54. IPCC AR5 WG3 SPM, Figure SPM.4 and Table SPM.1. These requirements are based on the median pathways presented for each scenario.

55. Kevin Anderson. "Duality in climate science." *Nature Geoscience* 8 (2015). [online]. doi:10.1038/ngeo2559.

56. IPCC AR5 synthesis report SPM. The total estimated budget from 1870 onward is 790 [695 to 858] GtC, and 520 GtC was emitted up to 2011.

57. Anderson. "Duality in climate science."

58. In my survey of Earth scientists, when asked "Do you believe that we will stay below the 2 degree Celsius threshold?" 92% said "no" while 8% said "yes" (60 responses).

59. Christophe McGlade and Paul Ekins. "The geographical distribution of fossil fuels unused when limiting global warming to 2°C." *Nature* 517 (2015). [online]. doi:10.1038/nature14016.

60. I looked at current costs of wind, solar voltaic, concentrating solar trough, and nuclear energy, and electrical storage; made some basic assumptions about energy storage; and assumed that as a nation we can cut our electricity use in half. This led to an estimate that electrical decarbonization would cost about $1 trillion. This is about $3,000 per person, or alternatively the combined net worth of the twenty or so richest Americans. By comparison, by 2015, the "war on terror" had cost US taxpayers $1.7 trillion: Niall McCarthy. "The war on terror has cost taxpayers $1.7 trillion." *Forbes*, February 3 2015. [online]. forbes.com/sites/niallmccarthy/2015/02/03/the-war-on-terror-has-cost-taxpayers-1-7-trillion-infographic/#75a6f1255cfo.

61. Three quarters of this was from hydropower. World Energy Council. "Variable renewable energy sources integration in electricity systems 2016: How to get it right." (2016). [online]. worldenergy.org/publications/2016/variable-renewable-energy-sources-integration-in-electricity-systems-2016-how-to-get-it-right/.

62. Kanyakrit Vongkiatkajorn. "California just took a huge step in the fight against climate change." *Mother Jones*, September 8, 2016. [online]. motherjones.com/environment/2016/09/california-passes-sb-32-groundbreaking-climate-legislation.

63. It's important to realize that no scientist will say he or she is 100% certain about any future outcome, ever. No one can be 100% certain about any event in the future. There should be a word in our language which scientists can use to mean "certain beyond any reasonable doubt" and which non-scientists hear as "absolutely certain." I've chosen to use the word "unequivocal" for this purpose.

64. Keynyn Brysse et al. "Climate change prediction: Erring on the side of least drama?" *Global Environmental Change* 23(1) (2013). [online]. doi:10.1016/j.gloenvcha.2012.10.008.

Chapter 5: Growth Always Ends

1. With the spacecraft New Horizon's 2015 flyby of Pluto, NASA has now explored every world in our solar system. One thing we've learned is that every planet in our system except Earth is incredibly hostile to human life. Another thing we've learned is that space travel is very difficult: NASA workers hold their breath at every critical juncture of a mission (e.g., launch, deployment, orbit insertion, landing) as there are so many ways for the mission to fail. In my opinion, it's likely that we'll identify a plausibly Earth-like planet around another star within the lifetimes of humans alive today. However, even sending just a handful of people to this world would be far beyond our current technology—we can't even run a Biosphere 2 here on Earth. For a good discussion of the immense difficulties of interstellar travel, see: Tom Murphy. "Why Not Space?" *Do The Math* blog, October 12, 2011. [online]. physics.ucsd.edu/do-the-math/2011/10/why-not-space/.

2. I heard a basic version of this story from Chris Martenson, who heard it from Albert Allen Bartlett.

3. The stadium is a cylinder with 207,000 square meters of cross-sectional area and 60 meters of height.

4. We assume 0.05 milliliter of water per drop.

5. Fortunately, one of Bond's hobbies is freediving. He lowers his heart rate, and using a minimum of motion, he finishes escaping from the handcuffs and swims up to air. Transformed by the experience, he turns in his license to kill and becomes a first-rate high school math teacher.

6. For math people, here's why: the exponential function can be written $y/y_0 = e^{\ln(1+r)t} = (1 + r)^t$ where t is time, y_0 is the value of y at $t = 0$, and

r is the fractional growth rate per unit of time (i.e., $R = 100r$). After one doubling, when $t = t_d$, we have $2 = (1 + r)^{td}$. Taking the logarithm of both sides gives $t_d = \ln(2)/\ln(1 + r) \approx 0.693/r$ for small r (e.g., less than 0.15, or 15%).

7. Extrapolating this simple 2.2% growth per year into the past agrees reasonably well with actual historical concentrations, which were about 315 ppm CO_2 in 1960 and 350 ppm CO_2 in 1990. The minor discrepancies arise because the historical growth rate used to be less than 2.2% per year. But in the early stages of growth, inaccuracies due to the growth rate are small compared to the 280 ppm baseline.

8. Data are from: US Census Bureau. "World Population: Historical Estimates of World Population." International Data Base, revised September 27, 2016. [online]. census.gov/population/international/data/worldpop/table_history.php. Where higher and lower estimates were provided, I have taken the mean.

9. Additional data and projections into the future are from: United Nations, Population Division. *World Population Prospects: The 2012 Revision*, medium fertility variant. [online]. esa.un.org/unpd/wpp/Publications/.

10. Data from: US Census Bureau. "World Population: Total Midyear Population for the World, 1950–2050." International Data Base, revised September 27, 2016. [online.] census.gov/population/international/data/worldpop/table_population.php.

11. Elina Pradhan. "The relationship between women's education and fertility." World Economic Forum, November 27, 2015. [online]. weforum.org/agenda/2015/11/the-relationship-between-womens-education-and-fertility.

12. This would be precisely true if population growth was equal across individual emissions levels, but population growth is higher in poorer countries which have lower *per capita* emissions. This implies that growth in *per capita* emissions may be a larger driver of global emissions than population growth.

13. With an 80% confidence interval of 9.6 billion to 10.0 billion.

14. With an 80% confidence interval of 10.0 billion to 12.5 billion. This projection should be taken with a grain of salt, however. For example, in 1951, the UN predicted a world population of 3 billion in 1980; but the actual was 4.4 billion, 50% higher than predicted.

15. United Nations, Population Division. *World Population Prospects: The 2015 Revision*. [online]. esa.un.org/unpd/wpp/Publications/.

16. Some people frame this as an either/or proposition, with proponents of capitalism calling for population control and proponents of socialism calling for a more equitable division of resources (with no need

for population control). At this time, however, humans are emitting too much CO_2 into the atmosphere, and the fact is that this is a function of both our modes of resource use *and* our population.

17. Thomas J. Espenshade et al. "The surprising global variation in replacement fertility." *Population Research and Policy Review* 22(5/6) (2003). [online]. doi:10.1023/B:POPU.0000020882.29684.8e.

18. Kristin Park. "Stigma Management among the Voluntarily Childless." *Sociological Perspectives* 45(1) (2002). [online]. jstor.org/stable/10.1525 /sop.2002.45.1.21.

19. Deepak K. Ray et al. "Recent patterns of crop yield growth and stagnation." *Nature Communications* 3 (2012). [online]. doi:10.1038 /ncomms2296.

20. Norman Borlaug. "The Green Revolution, Peace, and Humanity." Nobel lecture, December 11, 1970. [online]. nobelprize.org/nobel _prizes/peace/laureates/1970/borlaug-lecture.html.

21. Staple crops such as wheat, rice, corn, and soybeans were bred for dwarfing (dwarf plants divert a higher fraction of their energy into carbohydrates), shorter maturation times, disease resistance, responsiveness to fertilizers and irrigation, insensitivity to herbicides, and insensitivity to day length (allowing for success in a wider range of latitudes, and allowing farmers in some regions to grow in two seasons instead of only one).

22. Engineers are modifying both plants and animals, although GMO animals have yet to take off. GMO animals include fast-growing salmon (with inserted eel genes), pigs with larger butts (with an introduced mutation), "web-spinning" goats (with inserted spider genes), and glow in the dark cats (with inserted jellyfish genes).

23. E. T. Lammerts van Bueren et al. "The need to breed crop varieties suitable for organic farming, using wheat, tomato and broccoli as examples: A review." *NJAS—Wageningen Journal of Life Sciences* 58(3–4) (2011). [online]. doi:10.1016/j.njas.2010.04.001.

24. Rhys E. Green et al. "Farming and the fate of wild nature." *Science* 307(5709) (2005). [online]. doi:10.1126/science.1106049.

25. Nathaneal Johnson. "Do industrial agricultural methods actually yield more food per acre than organic ones?" *Grist*, October 14, 2015. [online]. grist.org/food/do-industrial-agricultural-methods-actually -yield-more-food-per-acre-than-organic-ones/.

26. US Environmental Protection Agency. "Nutrient Pollution: The Problem." [online]. epa.gov/nutrientpollution/problem.

27. Charles R. Fink et al. "Nitrogen fertilizer: Retrospect and prospect." *Publications of the National Academy of Sciences* 96(4) (1999). [online]. doi:0.1073/pnas.96.4.1175.

28. Robert W. Howarth. "Coastal nitrogen pollution: A review of sources and trends globally and regionally." *Harmful Algae* 8(1) (2008). [online]. dx.doi.org/10.1016/j.hal.2008.08.015.

29. Matthew Hora and Judy Tick. *From Farm to Table: Making the Connection in the Mid-Atlantic Food System.* Capital Area Food Bank of Washington DC, 2001. [online]. openlibrary.org/books/OL11779852 M/From_farm_to_table.

30. Joan Dye Gussow. *Chicken Little, tomato sauce and agriculture: Who will produce tomorrow's food?* The Bootstrap Press, 1991. [online]. worldcat.org/title/chicken-little-tomato-sauce-and-agriculture-who -will-produce-tomorrows-food/oclc/23583327.This energy ratio has been growing, so today's ratio is likely to be significantly higher.

31. Ibid.

32. Data from the IMFund, PFOOD, and PNRG price indices: International Monetary Fund. "IMF Primary Commodity Prices." [online]. imf.org/external/np/res/commod/index.aspx.

33. Suppose you have a male and a female mouse in a large cage with water and everything else mice need to be happy. Every morning, you give the mice a pound of food; every evening, you remove the leftover food. The mice have babies, the babies have babies, and the population grows. Eventually the population reaches a certain number of mice, call it N (the carrying capacity, in this case the number of mice that a pound of food per day can support), and over time you observe that it stays pretty close to that same number. You also notice that the mice eat all the food by the end of the day. The population has reached equilibrium. What happens if you double the food? The population will climb until there are $2N$ mice and settle at that new equilibrium. The Green Revolution is evidence that human populations also follow this basic law of ecology. See: R. L. Strecker and J. T. Emlen. "Regulatory mechanisms in house-mouse populations: The effect of limited food supply on a confined population." *Ecology* 34(2) (1953). [online]. doi:10.2307/1930903.

    Interestingly, the mouse population will stop growing very suddenly because the mice abruptly stop reproducing when they reach carrying capacity. The biological mechanism for this involves physiological changes to their reproductive organs.

34. Relative to demand in the year 2000: Green. "Farming and the fate of wild nature."

35. UN FAO. "The State of Food Insecurity in the World." 2009 and 2015 reports. [online]. ao.org/hunger/en/. Note that the FAO has been accused of revising past estimates in order to demonstrate positive progress on world hunger: see Martín Caparrós. "Counting the

hungry." *New York Times*, September 27, 2014. [online]. nytimes.com /2014/09/28/opinion/sunday/counting-the-hungry.html.

36. Patricio Grassini et al. "Distinguishing between yield advances and yield plateaus in historical crop production trends." *Nature Communications* 4 (2013). [online]. doi:10.1038/ncomms3918.

37. See e.g., Prabhu L. Pingali. "Green Revolution: Impacts, limits, and the path ahead." *Publications of the National Academy of Sciences* 109(31) (2012). [online]. doi:10.1073/pnas.0912953109.

38. Michael P. Russelle et al. "Reconsidering Integrated Crop: Livestock Systems in North America." *Agronomy Journal* 99(2) (2006). [online]. doi:10.2134/agronj2006.0139.

39. Brenda B. Lin. "Resilience in Agriculture through Crop Diversification: Adaptive Management for Environmental Change." *BioScience* 61(3) (2011). [online]. doi:10.1525/bio.2011.61.3.4.

40. My favorite anecdote about the modern reduction in crop bio-diversity is told by Jon Jondai. See his short video talk: John Jandai. "A personal story on seed saving." YouTube, 2011. [online]. youtube .com/watch?v=3BweruD8RyI.

41. Tom Gleeson et al. "Water balance of global aquifers revealed by groundwater footprint." *Nature* 488 (2012). [online]. doi:10.1038 /nature11295.

42. David R. Steward et al. "Tapping unsustainable groundwater stores for agricultural production in the High Plains Aquifer of Kansas, projections to 2110." *Publications of the National Academy of Sciences* 110(37) (2013). [online]. doi:10.1073/pnas.1220351110.

43. This could be seen as an opportunity. For example, we could choose to let the Ogallala grassland return to native prairie with grazing bison. Meriwether Lewis recorded his impression of the bison on numerous occasions in his journal. For example, on September 17, 1804, bound toward the Pacific in the land that is now South Dakota, he wrote, "This senery already rich pleasing and beautiful was still far-ther hightened by immence herds of Buffaloe deer Elk and Antelopes which we saw in every direction feeding on the hills and plains." On the way back, on August 29, 1806, he wrote, "I assended to the high Country.... From this eminance I had a view of a greater number of buffalow than I had ever Seen before at one time. I must have Seen near 20,000 of those animals feeding on this plain." Source: Dis-covering Lewis and Clark website. "Bison in the Journals." [online]. lewis-clark.org/article/443.

This native system of production requires no fossil water, fossil fuels, chemicals, tilling, or indeed human interventions of any kind.

Considering that this system was fine-tuned by nature over evolutionary time, and given the anecdotal reports of pre-agricultural abundance, it's possible that the prairie/bison system could yield more
meat per acre than our current corn/cattle system, although there has
been little research into this question. Ideally vast swaths of grassland
would be returned to the commons, owned by all. Implementing
this ideal version of the vison would require transforming some
of society's deepest tenets about agriculture and land ownership.
However, if productivity and profitability of the prairie/bison system
were indeed higher than the corn/cattle system in certain ecosystems,
a middle path would be possible. Consideration of these systems
should include fixing cost externalities in fossil fuel and fossil water
that effectively subsidize the corn/cattle system.

44. Predicted in the medium warming scenario presented in California
Climate Change Center. *Our Changing Climate: Assessing the Risks to
California*. Document # CEC-500-2006-077, 2006. [online]. meteora
.ucsd.edu/cap/pdffiles/CA_climate_Scenarios.pdf.

45. Benjamin I. Cook et al. "Unprecedented 21st century drought risk in
the American Southwest and Central Plains." *Science Advances* 1(1)
(2015). [online]. doi:10.1126/sciadv.1400082. This paper predicts that
droughts will get worse even if humans choose to mitigate global
warming, but they will be worse still if we choose not to.

46. UN FAO. *The State of the World's Land and Water Resources for Food
and Agriculture: Managing Systems at Risk.*" FAO Summary Report,
2011. [online]. fao.org/nr/water/docs/SOLAW_EX_SUMM_WEB
_EN.pdf.

47. Nigel Hunt and Sarah McFarlane. "'Peak soil' threatens future
global food security." *Reuters*, July 17, 2014. [online]. reuters.com
/article/us-peaksoil-agriculture-idUSKBN0FM1HC20140717. The
article cites a 30% yield decrease by 2050, and attributes this prediction to John Crawford, a soil scientist affiliated with Rothamsted
Research.

48. "In just one teaspoon of agricultural soil there can be one hundred
million to one billion bacteria, six to nine feet of fungal strands put
end to end, several thousand flagellates and amoeba, one to several
hundred ciliates, hundreds of nematodes, up to one hundred tiny
soil insects, and five or more earthworms. These organisms are
essential for healthy growth of your plants." S. Tianna DuPont. *Soil
quality: Introduction to soils*. Penn State College Extension, 2012, p. 6.
[online]. extension.psu.edu/business/start-farming/soils-and-soil
-management/soil-quality/extension_publication_file.

49. John W. Crawford et al. "Microbial diversity affects self-organization of the soil-microbe system with consequences for function." *Journal of the Royal Society Interface* 9(71) (2012). [online]. doi:10.1098/rsif .2011.0679.

50. Hunt and McFarlane. "'Peak soil.'"

51. Erica Goode. "Farmers put down the plow for more productive soil." *New York Times*, March 9, 2015. Much of this no-till farming uses herbicides to kill cover crops. However, cover crops can readily be killed mechanically, making organic no-till feasible: see Rodale Institute. "Our Work: Organic No-Till." [online]. rodaleinstitute.org /our-work/organic-no-till/.

52. Each year humans destroy 13 million hectares of forest (90,000 acres per day), most of it in tropical rainforests, but some forest regenerates; the net annual loss is 5.2 million hectares. (These figures are annual means between 2000 and 2010.) US FAO. *State of the World's Forests, 2011.* Rome. [online]. fao.org/docrep/013/i2000e/i2000e00.htm.

53. IPCC AR5 WG2, Chapter 7.

54. David B. Lobell and Christopher B. Field. "Global scale climate-crop yield relationships and the impacts of recent warming." *Environmental Research Letters* 2(1) (2007). [online]. doi:10.1088/1748-9326/2/1 /014002.

55. Andrew E. Kramer. "Russia, crippled by drought, bans grain exports." *New York Times*, August 5, 2010. [online]. nytimes.com/2010/08/06 /world/europe/06russia.html.

56. Koh Iba. "Acclimative response to temperature stress in higher plants: Approaches of gene engineering for temperature tolerance." *Annual Review of Plant Biology* 53 (2001). doi:10.1146/annurev.arplant .53.100201.160729.

57. Daniel P. Bebber et al. "Crop pests and pathogens move polewards in a warming world." *Nature Climate Change* 3 (2013). [online]. doi:10.1038/nclimate1990.

58. Samuel S. Myers et al. "Increasing CO_2 threatens human nutrition." *Nature* 510 (2014). [online]. doi:10.1038/nature13179.

59. Lewis H. Ziska et al. "Rising atmospheric CO_2 is reducing the protein concentration of a floral pollen source essential for North American bees." *Proceedings of the Royal Society B* 283(1828) (2016). [online]. doi:10.1098/rspb.2016.0414.

60. Linda O. Mearns et al. "Effect of changes in interannual climatic variability on CERES-wheat yields: Sensitivity and 2xCO$_2$ general circulation model studies." *Agricultural and Forest Meteorology* 62(3&4) (1992). [online]. doi:10.1016/0168-1923(92)90013-T.

61. Stephen P. Long et al. "Food for thought: Lower-than-expected crop

yield stimulation with rising CO_2 concentrations," *Science* 312 (2006). [online]. doi:10.1126/science.1114722.

62. Ibid.

63. H. Charles J. Godfray. et al. "Food security: The challenge of feeding 9 billion people." *Science* 327 (2010). [online]. doi:0.1126/science .1185383.

64. Deepak K. Ray et al. "Yield Trends Are Insufficient to Double Global Crop Production by 2050." *PLOS One* 8(6) (2013). [online]. doi:10.1371/journal.pone.0066428.

65. Rabah Arezki and Markus Brückner. "Food prices and political instability." International Monetary Fund Working Paper #WP/11/62, 2011. [online]. imf.org/external/pubs/ft/wp/2011/wp1162.pdf.

66. Ben Laffin and Megan Specia. "Venezuela Gripped by Hunger and Riots." *New York Times Video*, June 21, 2016. [online]. nytimes.com /video/world/americas/100000004485562/venezuela-gripped-by -hunger-and-riots.html.

67. McGlade and Ekins. "The geographical distribution of fossil fuels," p. 190.

68. Donald W. Jones et al. "Oil price shocks and the macroeconomy: What has been learned since 1996." *Energy Journal* 25(2) (2004). [online]. doi:10.2307/41323029.

69. Kevin Drum. "Peak oil and the great recession." *Mother Jones*, October 19, 2011. [online]. motherjones.com/kevin-drum/2011/10/peak -oil-and-great-recession.

70. Art Berman. "Despite OPEC production cut, another year of low oil prices is likely." *Forbes*, January 9, 2017. [online]. forbes.com/sites/ arthurberman/2017/01/09/the-opec-oil-production-cut-another-year -of-lower-oil-prices.

71. You can see them for yourself in satellite images on an online map tool. For example, enter the coordinates "40N, 109.33W" in the search bar, and switch to the satellite view. Those strangely repetitive structures are fracking wells. Now hold on to your seat, and zoom out a few times: welcome to the Matrix. There are many similar areas in other parts of the US.

72. Data: US Energy Information Administration. "U.S. Field Production of Crude Oil" and "Total Petroleum and Other Liquids Production." [online]. eia.gov.

73. A consequence of the aggregation of many individual wells and the central limit theorem from statistics.

74. The functional form for the sum of two Hubbert curves is:

$$y(t) = \frac{\beta_1}{1 + \cosh(\beta_2[t - \beta_3])} + \frac{\beta_4}{1 + \cosh(\beta_5[t - \beta_6])}$$

75. J. David Hughes. *Drilling deeper: A reality check on U.S. government forecasts for a lasting tight oil and shale gas boom.* Post Carbon Institute, October 2014. [online]. postcarbon.org/wp-content/uploads /2014/10/Drilling-Deeper_FULL.pdf. At the time of writing, this was the most thorough report available; however, some might claim that the Post Carbon Institute is biased towards predicting an early peak.

76. US Energy Information Administration. *U.S. Crude Oil Production to 2025: Updated Projection of Crude Types.* May 25, 2015, p. 1. [online]. eia.gov/analysis/petroleum/crudetypes/pdf/crudetypes.pdf.

77. G. Maggio and G. Cacciola. "When will oil, natural gas, and coal peak?" *Fuel* 98 (2012). [online]. doi:10.1016/j.fuel.2012.03.021.

78. Cutler J. Cleveland. "Energy and the US economy: A biophysical perspective." *Science* 225(4665) (1984). [online]. doi:10.1126/science.225 .4665.890.

79. Nathan Gagnon et al. "A preliminary investigation of the energy return on energy investment for global oil and gas production." *Energies* 2(3) (2009). [online]. doi:10.3390/en20300490.

80. Charles A.S. Hall et al. "EROI of different fuels and the implications for society." *Energy Policy* 64 (January 2014). [online]. doi:10.1016/j .enpol.2013.05.049.

81. Ibid.

82. Ibid. and references contained therein. Note that it's difficult to make EROEI estimates, and I've rounded the published values to one significant figure to reflect this uncertainty (Hall "EROI of different fuels" does not provide uncertainty estimates).

83. Mean world GDP growth rates for the 1960s, 1970s, 1980s, 1990s, 2000s, and 2010 to 2015 were 5.52%, 4.11%, 3.07%, 2.66%, 2.86%, and 2.95%. Mean US GDP growth rates for the same periods were 4.66%, 3.54%, 3.14%, 3.23%, 1.82%, and 2.17%. Data from: World Bank. "GDP Growth (annual %)" [online]. data.worldbank.org/indicator/NY .GDP.MKTP.KD.ZG.

84. For example, section 309a of the California Corporations Code states: "A director shall perform the duties of a director...in good faith, in a manner such director believes to be in the best interests of the corporation and its shareholders." [online]. codes.findlaw.com/ca /corporations-code/corp-sect-309.html. Your jurisdiction no doubt has a similar statute.

85. Citizens United v. Federal Election Commission. 558 U.S. 310 (2010). [online]. supremecourt.gov/opinions/09pdf/08-205.pdf.

86. American Legislative Exchange Council. [online]. alec.org.

87. Fortune. *Global 500.* [online]. fortune.com/global500/2015/. Among the ten largest corporations in 2015, six of the top ten rankings were

held by petroleum companies; two other spots in the top ten were held by automotive companies.

88. Naomi Oreskes and Erik Conway. *Merchants of Doubt: How a Handful of Scientists Obscured the Truth on Issues from Tobacco Smoke to Global Warming*. Bloomsbury, 2010.

89. The United Nations Framework Convention on Climate Change. Article 3: Principles. [online]. unfccc.int/cop4/conv/conv_005.htm.

90. Money Network Alliance. "The money system requires continual growth." [online]. monneta.org/en/the-money-system-requires -continual-growth/.

91. First, increasing the efficiency of our production systems and gadgets can only take us so far: even a perfectly efficient microwave oven (or replicator, for that matter) still needs to source at least 240 grams of clean water and 80,000 Joules of energy for that cup of tea (Earl Grey, hot). And if real economic growth were to somehow continue exponentially while energy and resource use remain fixed, after a few doublings of the economy, we'd reach an absurdity in which all the resources and energy in the world could be purchased with a single worker's daily wage. For the full *reductio ad absurdum* argument against decoupling, see Tom Murphy. "Can Economic Growth Last?" *Do the Math* blog, July 14, 2011. [online]. physics.ucsd.edu/do-the -math/2011/07/can-economic-growth-last/.

92. According to the World Wildlife Fund, populations of vertebrate species have dropped by 52% on average since 1970: World Wildlife Fund. *Living Planet Report 2014*. [online]. worldwildlife.org/pages /living-planet-report-2014.

93. Jurriaan M. De Vos et al. "Estimating the normal background rate of species extinction." *Conservation Biology* 29 (2015). [online]. doi:10.1111/cobi.12380.

94. Vaclav Smil. *The Earth's Biosphere: Evolution, Dynamics, and Change*. MIT Press, 2003. According to this source, human flesh accounts for 26% of the land vertebrate biomass total, while livestock flesh accounts for 71%. Note that these numbers are based on data from 1990 or earlier, when there were far fewer humans and far more wild animals, so today's situation is most probably even more unbalanced.

95. Brian MacQuarrie. "Ticks devastate Maine, N.H. moose popula- tions." *Boston Globe*, January 13, 2017. [online]. bostonglobe.com /metro/2017/01/13/winter-ticks-exact-heavy-toll-new-england-moose /PmpQ3QAHm9C1imAxkzMhDM/story.html.

96. Alejandro Estrada et al. "Impending extinction crisis of the world's primates: Why primates matter." *Science Advances* 3(1) (2017). [on- line]. doi:10.1126/sciadv.1600946.

97. Jared Diamond. *Collapse: How Societies Choose to Fail or Succeed.* Penguin, 2004.

98. Mathis Wackernagel et al. "Tracking the ecological overshoot of the human economy." *Publications of the National Academy of Sciences* 99(14) (2002). [online]. doi:10.1073/pnas.142033699.

99. Gretchen C. Daily et al. "Optimum human population size." *Population and Environment* 15(6) (1994). [online]. dieoff.org/page99.htm.

100. Christian J. Peters et al. "Carrying capacity of U.S. agricultural land: Ten diet scenarios." *Elementa: Science of the Anthropocene* 4(116) (2016). [online]. doi:10.12952/journal.elementa.000116. Interestingly, according to this study, the planet could actually support more dairy-eating vegetarians than vegans, because dairy animals can eat grass in regions too arid to farm.

101. UN Food and Agriculture Organization. "Key facts on food loss and waste you should know!" [online]. fao.org/save-food/resources /keyfindings/en/.

102. Jared Diamond. "The worst mistake in the history of the human race." *Discover Magazine*, May 1987. [online]. discovermagazine.com/1987 /may/02-the-worst-mistake-in-the-history-of-the-human-race. Diamond provides evidence that the following miseries resulted from switching to agriculture some 10,000 years ago: extended work hours; class systems; oppression of women; increased incidence of parasites and disease; increased risk of famine; malnutrition; and increased warfare. One could argue that agriculture also led to chattel slavery, and in Chapter 5, I've argued that it has led to overpopulation and global warming, as well. And this is just from the human perspective. For most nonhuman species (rats and wheat being two exceptions), agriculture has meant nothing but death.

103. Daniel Quinn. *The Story of B.* Bantam Books, 1996.

Chapter 6: Our Mindset

1. The theory of evolution, which challenged the Christian creation myth, provides an example of this.

2. Roland Barthes. *Mythologies*, trans. Annette Lavers. Hill and Wang, 1972, pp. 142–143.

3. Meadows. "Places to Intervene in a system."

4. By "we" I mean those of us operating through the myth of progress. This seems to include the vast majority of people in industrial society, cutting across economic and racial strata.

5. For this image, I thank John Michael Greer, whose eye-opening blog post "Which Way to Heaven?" was an influence for this chapter:

John Michael Greer. "Which Way to Heaven?" *The Archdruid Report*, September 25, 2013. [online]. thearchdruidreport.blogspot.ca/2013/09 /which-way-to-heaven.html.

6. Such is the case with exoplanetary astronomy. There are no exoplanet deniers; or if there are, they stay pretty quiet.

7. As we saw in Chapter 3, although the estimated quantities have uncertainties (and always will), and although there are still many details to pursue, the basic existence of a human-caused global radiative energy imbalance is as clear as anything in science can be.

8. Speaking on *Frontline*, Dr. Arjun Srinivasan, Associate Director at the Centers for Disease Control and Prevention, said: "For a long time, there have been newspaper stories and covers of magazines that talked about 'The end of antibiotics, question mark?' Well, now I would say you can change the title to 'The end of antibiotics, period.' We're here. We're in the post-antibiotic era." Sarah Childress. "Dr. Arjun Srinivasan: We've Reached 'The End of Antibiotics, Period.'" *Frontline*, October 22, 2013. [online]. pbs.org/wgbh/frontline/article/dr-arjun -srinivasan-weve-reached-the-end-of-antibiotics-period/.

9. See Robin McKie. "Millions at risk as deadly fungal infections acquire drug resistance." *Guardian*, August 27, 2016. [online]. theguardian .com/society/2016/aug/27/millions-at-risk-as-deadly-fungal -infections-acquire-drug-resistance.

10. Kenneth J. Loceya and Jay T. Lennona. "Scaling laws predict global microbial diversity." *Publications of the National Academy of Sciences* 113(21) (2016). [online]. doi:10.1073/pnas.1521291113.

11. Aboriginal Culture. "Religion and Ceremony." [online]. aboriginalculture.com.au/religion.shtml.

12. Daniel Kahneman. *Thinking, Fast and Slow*. Farrar, Straus & Giroux, 2013.

13. Solomon E. Asch. "Opinions and social pressure." *Scientific American* 193(5) (1955). [online]. scientificamerican.com/article/opinions-and -social-pressure/.

14. Overall, subjects conformed and gave the wrong answer a shocking 37% of the time.

15. See Chapter 11 for discussion of how this habit operates at the level of physical sensation.

Chapter 7: Trailheads into the Wilderness

1. According to the US Centers for Disease Control and Prevention, depression afflicts more than 26% of the US adult population; and by 2020 will be the second leading cause of disability throughout

the world, trailing only coronary heart disease: US CDC. "Mental Health Basics." [online]. cdc.gov/mentalhealth/basics.htm. Drug overdose deaths continue to rise: US CDC. "Drug overdose deaths in the United States continue to increase in 2015." [online]. cdc.gov /drugoverdose/epidemic/.

2. Palm oil is used in products such as buttery spreads, crackers, instant noodles, cosmetics, and soaps; it sometimes hides under other names in ingredient lists, such as stearic acid, sodium lauryl/laureth sulfate, and cetyl alcohol: Lael Goodman. "How Many Products with Palm Oil Do I Use in a Day?" Union of Concerned Scientists Blog, April 3, 2014. [online]. blog.ucsusa.org/lael-goodman/how-many-products -with-palm-oil-do-i-use-in-a-day. Half of all packaged products in the supermarket: Rosie Spinks. "Why does palm oil still dominate the supermarket shelves?" *The Guardian*, December 17, 2014. [online]. theguardian.com/sustainable-business/2014/dec/17/palm-oil -sustainability-developing-countries.

   Sustainability of palm oil is a complicated question: on a given patch of land, you can either have rainforest or oil palms, but not both. But this is true of other crops in other biomes, as well.

3. Vipassana Meditation. Dhamma.org. [online]. dhamma.org

4. Citizens' Climate Lobby is an international volunteer organization dedicated to creating political will for real climate action. [online]. citizensclimatelobby.org.

Chapter 8: Like to Bike

1. At that time, we still made a Christmas pilgrimage to Illinois, loading the family onto an airplane every year. We've stopped doing this and are no worse off; see Chapter 10.

2. Notes from a recent long ride to an event on the other side of Los Angeles: "I could have taken public transit, but I didn't want to miss the bike ride. It was powerful—I felt powerful. Thought it was out of my comfort zone a little but then realized it was well within. And I made all these connections. To places I'd driven to. To the layout of the city. To the city's efforts to make better bike paths. To the neigh-borhoods. To the sky. To my own amazing body. It felt really good, in my body. Also in my mind and my spirit."

3. Worse car commutes are correlated with higher divorce rates. Annie Lowrey. "Your Commute Is Killing You." May 26, 2011, slate.com. [online]. slate.com/articles/business/moneybox/2011/05/your _commute_is_killing_you.html.

4. Scientific source: Ingrid J.M. Hendriksen et al. "The association

between commuter cycling and sickness absence." *Preventive Medicine* 51(2) (2010). [online]. doi:10.1016/j.ypmed.2010.05.007.

   Anecdotal source: For several years before I started biking, I suffered through a series of sinus infections, about one per year. Since getting back on my bike in 2009, I've only had one (during a period when I was biking less), and I've hardly been sick at all. Biking makes me feel generally great.

5. Biking saves our family more than $2,000 per year at the 2016 IRS mileage rate of $0.54 per mile, after accounting for the $150 or so per year we spend on bike maintenance. This presumes that one bike mile ridden replaces one car mile ridden (it actually replaces more, because on a bike I plan my trips carefully), and it doesn't include indirect savings such as not needing a gym membership or seldom getting sick.

6. For every hour I spend keeping my bike in good shape, I likely spend 20 or more keeping my old car, Maeby, growling along. Maeby is much more complex, and can break in many more ways.

7. Herb Weisbaum. "What's the life expectancy of my car?" *NBC News*, March 28, 2006. [online]. nbcnews.com/id/12040753/ns/business -consumer_news/t/whats-life-expectancy-my-car/.

8. Errands on bikes are bound to be local, simply because it requires an investment of personal energy.

9. For example, shifting from a car commute to a bike commute.

10. Jeroen Johan de Hartog et al. "Do the health benefits of cycling outweigh the risks?" *Environmental Health Perspectives* 118 (2010). doi:10.1289/ehp.0901747. Estimates are based on life table calculations applied to a population of 500,000 people aged 18–64.

   Note that in a small number of extremely polluted cities, such as Delhi, India, breathing the air can outweigh the benefits of the exercise: Nick Van Mead. "Tipping point: Revealing the cities where exercise does more harm than good." *The Guardian*, February 13, 2007. [online]. theguardian.com/cities/2017/feb/13/tipping-point -cities-exercise-more-harm-than-good. I find this horrifying, and a great argument for clean air policies such as carbon fee and dividend (Chapter 14).

11. How does this result translate to US roads? De Hartog et al. used 2008 accident statistics from the Netherlands, where the mortality per passenger mile (in the 20 to 70 year age group) was 4.3 times higher for biking than for driving. Unfortunately, biking is more dangerous in the US, where the mortality per passenger mile was about 7 times higher for biking than for driving (there were 11 car deaths per

billion passenger miles in 2011; and 680 bicycle deaths [2011] over 9 billion bicycle miles): US DOT, Federal Highway Administration. *National Household Travel Survey 2009.* [online]. nhts.ornl.gov/intro-duction.shtml. Adjusting the factor of nine from de Hartog by the ratio 4.3/7 implies that biking in the US is only 6 times safer for one's overall health than driving. However, de Hartog et al. analyzes bicycle trips that occur in traffic 100% of the time. But about one-half of my commute is on a bike path through a nature preserve with no traffic, and the same is true for Sharon's commute. If this is typical, it might cut risk exposure approximately in half, bringing the overall safety factor in the US back up to about ten.

12. David Rojas-Rueda et al. "The health risks and benefits of cycling in urban environments compared with car use: Health impact assessment study." *British Medical Journal* (2011). [online]. doi:10.1136/bmj .d4521. In one year in Barcelona, 182,000 residents using the bike sharing system, compared with car users, experienced annual changes in mortality of 0.03 additional deaths from road traffic incidents, 0.13 additional deaths from air pollution, but avoided 12.46 deaths due to physical activity (a benefit to risk ratio of 77).

13. In 2011, 23% of people in the US who died riding bicycles were legally drunk: US DOT, National Highway Safety Traffic Administration. "Traffic Safety Facts, 2011 Data: Bicyclists and Other Cyclists." Doc# DOT HS 811 743 (April 2013), p. 4. [online]. crashstats.nhtsa.dot.gov /Api/Public/ViewPublication/811743. Those who avoid drinking and riding immediately gain a large additional safety margin. I also try to avoid riding when drunk drivers are more likely to be out.

14. A disproportionate number of bicycle accidents and fatalities involve the cyclist breaking rules of the road or otherwise violating common sense. For example, a 1992 study observing bikers in Palo Alto found that 15% of cyclists rode against traffic, and their risk of an accident was 3.6 times as high as those riding with traffic: Alan Wachtel and Diana Lewiston. "Risk Factors for Bicycle-Motor Vehicle Collisions at Intersections." *Institute of Transportation Engineers Journal* 64(9) (1994). [online]. bicyclinglife.com/Library/riskfactors.htm.

15. Such as getting crushed by a right-turning vehicle when a light turns green because you weren't far forward enough or far back enough: Michael Bluejay "Ten Ways Not to Get Hit—Collision Type #5: The Red Light of Death." Bicyclesafe website, updated May 2013. [online]. bicyclesafe.com.

16. I was unable to find reliable research quantifying the safety margin provided by a bike helmet. A simple thought experiment is enough

for me: in the unlikely event that I find myself flying headfirst towards concrete, I'd want a helmet on my head. However, it's better not to get hit in the first place.

17. Incredibly, only 15% of riders use lights at night: City of Boston. "Boston Bicycle Plan." Boston Transportation Department, 2001, p. 14. [online]. cityofboston.gov/transportation/accessboston/pdfs /bicycle_plan.pdf. Nearly ½ of US biking deaths occur after dark without bike lights, although only maybe 3% of rides happen after dark: City of Cambridge. "Bicycling Rules of the Road." Community Development Department, 2011. [online]. cambridgema.gov/cdd /transportation/gettingaroundcambridge/bybike/rulesoftheroad. For goodness sake, if you ride at night, use lights!

18. Some terminology of bike infrastructure: *bike paths* are car-free bicycle roads; *bike tracks* run alongside car roads but are physically separated from them; and *bike lanes* are space on car roads, demarcated only by a painted line, and are often situated between traffic and parked cars.

19. I restricted my counting to cars on my route, which didn't include the thousands of additional cars I saw on the freeway that parallels part of the route.

20. Brian McKenzie. "Modes Less Traveled: Bicycling and Walking to Work in the United States: 2008–2012." US Census Bureau (2014). [online]. census.gov/library/publications/2014/acs/acs-25.html.

21. Netherlands Ministry of Transport, Public Works and Water Management. "Bicycle Use in the Netherlands." *Cycling in the Netherlands,* 2009, section 1.1. [online]. fietsberaad.nl/library/repository /bestanden/CyclingintheNetherlands2009.pdf.

22. "Over the period 2005–2007 inhabitants of Amsterdam used their bikes on average 0.87 times a day, compared to 0.84 for their cars." CROW Fietsberaad. "Amsterdam: For the first time more transfers by bike than by car." News article, January 22, 2009. [online]. fietsberaad .nl/index.cfm?lang=en&section=Nieuws&mode=newsArticle&news Year=2009&repository=Amsterdam:+for+the+first+time+more +transfers+by+bike+than+by+car.

23. In the US, these priorities are baked into the metric that state departments of transportation use to design and evaluate roadways: Level of Service (LOS) means the rate of *cars* passing through an intersection.

24. Peter L. Jacobsen. "Safety in numbers: More walkers and bicyclists, safer walking and bicycling." *Injury Prevention* 9 (2003). [online]. doi:10.1136/ip.9.3.205. This study was prompted by the Pasadena City

Council in 1998 (which I naturally find interesting). The evidence for the safety in numbers effect is overwhelming, and there are many other studies.

25. At 11.3 kg CO_2 per gallon, which include upstream emissions; see Chapter 9.

26. See Constantine Samaras and Kyle Meisterling. "Life Cycle Assessment of Greenhouse Gas Emissions from Plug-in Hybrid Vehicles: Implications for Policy." *Environmental Science & Technology* 42(9) (2008). Supporting information. [online]. doi:10.1021/es702178s; Mike Berners-Lee and Duncan Clark. "What's the carbon footprint of…a new car?" *Guardian* Green Living blog, September 23, 2010. [online]. theguardian.com/environment/green-living-blog/2010/sep/23/carbon-footprint-new-car.

27. Jay Schwartz. "Calories burned biking one mile." Livestrong.com, November 2, 2013. [online]. livestrong.com/article/135430-calories-burned-biking-one-mile/. Yes, this does mean that Tour de France riders eat about four times as much as the spectators watching them.

28. This assumes 3,800 kcal of produced food and 2,100 kcal of consumed food per day, the difference being wasted food.

29. Using a ratio of typical car and bike weights, and assuming a 40,000-mile bike lifetime.

30. Considering only the embodied impact in the bike.

31. Michael Bluejay. "How to Not Get Hit by Cars: Important Lessons in Bicycle Safety." Bicycle Safe, May 2013. [online]. bicyclesafe.com.

32. Kurt Holzer. "Bike Law attorney Kurt Holzer makes a compelling case for the 'Idaho Stop.'" Bikelaw.com blog, January 27, 2016. [online]. bikelaw.com/2016/01/27/living-with-stop-as-yield-for-cyclists/.

Chapter 9: Leaving Fossil Fuel

1. In 2012, Bangladeshi per capita emissions were 1.2 tonnes CO_2e (1.0 tonnes CO_2e excluding land-use change), according to the World Resources Institute, CAIT Climate Data Explorer. [online]. cait.wri.org/historical.

2. There is one change I'd probably roll back, to a degree, if global warming suddenly, magically, disappeared: I'd fly occasionally, though far less than I did before I reduced. Still, overall I've gained more than I've sacrificed by giving up flying.

3. Charles Eisenstein discusses the story of separation and the importance of kindness in: Charles Eisenstein. *The More Beautiful World Our Hearts Know Is Possible*, 3rd ed. North Atlantic, 2013. Kindness is necessary but it's not sufficient: we also need to rapidly reduce our greenhouse gas emissions.

4. Here, I'll use the mass of CO_2 as opposed to the mass of carbon: 3.67 kg CO_2 contains 1 kg of carbon (the oxygen atoms add mass). In other parts of the book, I may give quantities as the mass of carbon (e.g., GtC in Chapter 3). Also, I'll round numbers to what I think is a reasonable degree of precision. In cases where I don't give uncertainty explicitly, the uncertainty estimate can be inferred from the numerical precision. For example, 120 implies more uncertainty (roughly 10%) than 121 (roughly 1%). However, I reserve the right to give an extra digit of precision if I feel that omitting it would bias a comparison or calculation.

Be aware that various analyses of greenhouse gas emissions use different methods and make different assumptions. Be careful to assess the assumptions underlying any given analysis. If these assumptions aren't clear, or if the underlying methodology or sources of information aren't clear, be wary of the conclusions.

5. Upstream emissions add an additional 28% [24%–31%] for gasoline, 20% [15%–25%] for diesel, 21% [17%–24%] for jet fuel, and 15% [9%–20%] for natural gas. See US EPA *Greenhouse Gas Emissions from the U.S. Transportation Sector: 1990–2003*. Office of Transportation and Air Quality, 2006, Table 14.1. [online]. nepis.epa.gov. These upstream emission estimates only consider CO_2 emissions; we will see below that methane emissions from natural gas production are also significant.

6. NO and NO_2 which form ozone, a greenhouse gas, in the upper troposphere.

7. The mean of the IPCC's estimates of the European average emission rate (short haul) and the rate for a 747 with 70% occupancy average over a 7,500-mile trip (long haul) is 0.30 kg CO_2 per coach passenger mile. However, the mean of the EPA's estimates for long and short flights is 0.21 kg CO_2 per coach passenger mile. I average the IPCC and EPA estimates and then factor in upstream emissions of 21%.

8. David J. Unger. "First-class ticket: More legroom, more emissions." *Christian Science Monitor*, June 17, 2013. [online]. csmonitor.com /Environment/Energy-Voices/2013/0617/First-class-ticket-more -legroom-more-emissions.

9. Christian Azar and Daniel J. A. Johansson. "Valuing the non-CO_2 climate impacts of aviation." *Climatic Change* 111 (2012). doi:10.1007 /s10584-011-0168-8.

10. Worldwatch Institute. *Vital Signs 2006–2007: The Trends That Are Shaping Our Future*. Norton, 2006, p. 68.

11. Or 2,500 kg CO_2e if we include non-CO_2 effects (via a multiplicative factor of 2.5).

12. Domestic and international carriers at all US airports tallied 1.3 trillion revenue passenger miles in 2015: US DOT, Bureau of Transportation Statistics. "Passengers—All Carriers—All Airports." [online]. transtats.bts.gov/Data_Elements.aspx?Data=1. US population in 2015: 321 million: US census. [online]. census.gov/popclock. This leads to a very rough estimate that ignores foreign travelers in the US, and US travelers in foreign countries.

13. I actually love the act of flying itself. As a boy, the National Air and Space Museum's brilliant 1976 film *To Fly!* utterly captivated me. I'd make my parents take me to see it each spring during our Easter road trip to Washington, DC to visit my grandparents. During college I learned how to fly sailplanes—small high-performance planes without engines that can soar on updrafts to heights of 50,000 feet— soloing, obtaining my license, and eventually wooing Sharon with soaring flight.

14. I discuss my transition to slow travel in Chapter 10.

15. US *per capita* vehicle miles of travel in 2013 was 9,400 miles: Chris McCahill. "Per capita VMT drops for ninth straight year: DOTs taking notice." State Smart Transportation Initiative, February 24 2014. [online]. ssti.us/2014/02/vmt-drops-ninth-year-dots-taking-notice/. The average fuel economy of cars on the road in 2013 was 21.6 miles per gallon: US DOT. "Average Fuel Efficiency of U.S. Light Duty Vehicles." Bureau of Transportation Statistics, Table 4-23. [online]. ita.dot.gov/bts/sites/rita.dot.gov.bts/files/publications/national _transportation_statistics/html/table_04_23.html.

16. This term originated in her poem, "Leave."

17. Burning a gallon of gas or diesel releases 8.9 and 10.2 kg CO_2, respectively: US Energy Information Agency. "Voluntary Reporting of Greenhouse Gases Program Fuel Emission Coefficients," Table 2. [online]. eia.gov/oiaf/1605/coefficients.html#tbl2. I've then added upstream emissions of 28% for gasoline and 20% for diesel (these are 2015 values; upstream emissions may increase over time).

18. The industrial agricultural processes used to grow this crop do result in fossil emissions (that is, CO_2 molecules made from carbon that had previously been locked safely underground). However, because I'm using a waste product, I don't count these emissions.

19. To calculate this figure: my commute was 6 miles each way, and at first, I rode it on a 35-mpg motorcycle.

20. Patrick McGeehan et al. "Beneath cities, a decaying tangle of gas pipes." *New York Times*, March 23, 2014. [online]. nytimes.com /2014/03/24/nyregion/beneath-cities-a-decaying-tangle-of-gas -pipes.html.

21. The composition of raw natural gas varies among gas fields, ranging between 70% to 90% methane, with the bulk of the remainder being ethane (an insignificant contributor to global warming), rounded out by small amounts of propane, butane, nitrogen, helium, CO_2, and various impurities. Refined natural gas—the gas delivered to your home—is nearly 100% methane: Natural Gas Supply Association. "Natural Gas: Background." [online]. naturalgas.org/overview /background.

22. Drew T. Shindell et al. "Improved attribution of climate forcing to emissions." *Science* 326 (2009) [online]. doi:10.1126/science.1174760. The uncertainty in these global warming potentials is 23%. Methane traps outgoing longwave radiation much more effectively than CO_2, but remains in the atmosphere for a shorter time. Therefore, comparing to CO_2 requires picking a time horizon. For more detail, see Chapter 3.

23. Most analyses underestimate methane by using a global warming potential of only 25 (choosing the 100-year time horizon and an outdated estimate of methane's impact). The choice of time horizon is subjective; I choose to use the average of the two conventional time horizons to avoid bias.

24. To estimate this: burning one therm of natural gas emits 5.3 kg CO_2, and burning an energy-equivalent amount of coal emits 10 kg CO_2. At a GWP of 65, 72 grams of methane has a climate impact equivalent to the difference, 4.7 kg CO_2. But the methane in a therm of natural gas has a mass of 2 kg, and 0.072 kg / 2 kg = 3.6%.

25. Robert W. Howarth et al. "Methane and the greenhouse-gas footprint of natural gas from shale formations." *Climatic Change* 106 (2011). [online]. doi:10.1007/s10584-011-0061-5.

26. Anna Karion et al. "Methane emissions estimate from airborne measurements over a western United States natural gas field." *Geophysical Research Letter* 40 (2013). [online]. doi:10.1002/grl.50811.

27. Interstate Natural Gas Association of America (INGAA). "Pipeline Fun Facts." Natural Gas Facts, INGAA website. [online]. ingaa.org /Topics/Pipelines101/PipelineFunFacts.aspx.

28. Here's how I estimated this. Roughly half of the natural gas production in the US comes from unconventional gas, most of which is recovered via fracking: American Petroleum Institute. "Facts About Shale Gas." [online]. api.org/oil-and-natural-gas/wells-to-consumer /exploration-and-production/hydraulic-fracturing/facts-about -shale-gas. So I estimate a total methane leakage rate of 5%. One kg of CH_4 combusts into 2.75 kg of CO_2; from this we can estimate that the methane in one therm of natural gas weighs about 2 kg; the 5%

leakage amounts to 100 grams. At a GWP of 65 times CO_2, this comes to 6.5 kg CO_2e per therm.

29. Residential natural gas use in the US in 2014 was 5.1 trillion cubic feet (52 billion therms): US Energy Information Administration. "Natural Gas Consumption by End Use." [online]. eia.gov/dnav/ng/ng_cons _sum_dcu_nus_a.htm.

30. A typical household of four would realize a natural gas reduction of about three tonnes CO_2e per year by installing a solar hot water heater.

      The average four-person household in the US uses about 70 gallons of hot water per day: Danny S. Parker et al. "Estimating Daily Domestic Hot-Water Use in North American Homes." *ASHRAE Transactions* 121(2) (2015). [online]. fsec.ucf.edu/en/publications/pdf /FSEC-PF-464-15.pdf.

      Heating this water requires about 230 therms of gas per year, assuming an efficient water heater (efficiency factor of 0.71): US Office of Energy Efficiency and Renewable Energy. "Energy Cost Calculator for Electric and Gas Water Heaters." [online]. energy.gov /eere/femp/energy-cost-calculator-electric-and-gas-water-heaters-0. Burning 230 therms of gas emits 3,000 kg CO_2e. Hot water energy use depends on climate as well as personal habits.

31. The system price tag was actually $9,000, but I received a $3,500 rebate from California and a 30% federal tax credit on the remainder. This seems like a lot of money just for hot water! Part of the problem is that the smallest available system was twice as big as what my family needs; it had an 80-gallon tank, whereas we'd been perfectly happy with a 40-gallon tank. It's also more complicated than necessary; it uses a pump, whereas a thermosiphon would be simpler. Perhaps there's commercial potential for a 40-gallon (or even 20-gallon) thermosiphon system. I looked into designing and building a DIY system, but it's quite difficult and I didn't have time.

32. Recall from Chapter 3 that N_2O is a powerful greenhouse gas. The GWP_{20} (global warming potential on a 20-year horizon) of 1 kg of N_2O is 270 kg CO_2e.

33. The majority of this, over 80%, comes from agricultural production itself; the rest comes from such things as transportation, processing, and packaging: Natasha Gilbert. "One-third of our greenhouse gas emissions come from agriculture." *Nature* News, October 31 2012. [online]. doi:10.1038/nature.2012.11708.

34. P. J. Gerber et al. *Tackling climate change through livestock: A global assessment of emissions and mitigation opportunities.* UN FAO, 2013.

[online]. fao.org/docrep/018/i3437e/i3437e.pdf. (Note that the claim made by the documentary film *Cowspiracy*, that 51% of global emissions are from livestock, is incorrect.)

35. Peter Scarborough et al. "Dietary greenhouse gas emissions of meat-eaters, fish-eaters, vegetarians and vegans in the UK." *Climatic Change* 125(2) (2014). [online]. doi: 10.1007/s10584-014-1169-1. A full ⅓ of food produced in the US is wasted, and these numbers include that waste. However, they assume a methane GWP of 25 and are therefore likely on the low side.

36. The Vegetarian Resource Group. "2016 National Poll." [online]. vrg.org/nutshell/Polls/2016_adults_veg.htm.

37. This 1,000 kg CO_2e does not include the methane emissions that I'm preventing by keeping this food out of the landfill, which amounts to an additional 1,000 kg CO_2e.

38. In 2014, the total US generation was 4.1e12 kWh, with coal and natural gas generating 39% and 27% respectively (the rest was nuclear, hydro, and renewables): US Energy Information Administration. "Electric Power Monthly." [online]. eia.gov/electricity/monthly/epm_table _grapher.cfm?t=epmt_1_1. This required 854 million short tons of coal and 88 billion therms of natural gas: US EIA. "Electric Power Monthly." Data for December 2015, February 2016, full report, tables 2.1.A and 2.4.A. [online]. eia.gov/electricity/monthly.

   Since a therm of natural gas produces 12.6 kg CO_2e (includes up-stream emissions and leakage), burning natural gas for electricity in 2014 produced 1.1 trillion kg CO_2e. The mean emissions from burning a short ton of coal is about 2,640 kg CO_2: US EIA. Carbon Dioxide Emission Factors for Coal." [online]. eia.gov/coal/production /quarterly/co2_article/co2.html; this value was for 1992, and I apply an 8% upstream emissions factor to get 2860 kg CO_2e per short ton: Paulina Jaramillo et al. "Comparative Life-Cycle Air Emissions of Coal, Domestic Natural Gas, LNG, and SNG for Electricity Generation." *Environmental Science and Technology* 41(17) (2007). [online]. doi:10.1021/es0630310. Burning coal for electricity in the US in 2014 thus generated 2.4 trillion kg CO_2e. Dividing the total emissions (3.5e12 kg CO_2e) by the total generated (4.1e12 kWh) and factoring in 6% transmission loss yields 0.9 kg CO_2e per kWh: US EIA FAQ. "How much electricity is lost in transmission and distribution in the United States?" [online]. eia.gov/tools/faqs/faq.cfm?id=105&t=3.

39. In 2015, the average US household used 10,812 kWh: US EIA FAQ. "How much electricity does an American home use?" [online]. eia.gov/tools/faqs/faq.cfm?id=97&t=3. There are 2.5 people per US

household: US Census Bureau, "Families and living arrangements,"
Table HH-6. Average Population Per Household and Family: 1940 to
Present. [online]. census.gov/hhes/families/data/households.html.

40. It costs an additional 3.5 cents per kWh.

41. Renewable electricity will soon be cheaper than fossil-fueled electricity, but as I've explained, this hasn't occurred yet for our household. Two things that will accelerate this important transition are
a revenue-neutral carbon fee (Chapter 14) and community choice
aggregation (Chapter 15).

42. Lindsay Wilson. "Shrink your product footprint." [online]. shrinkthat
footprint.com/shrink-your-product-footprint. To test this, consider
the Prius, which according to a detailed life cycle analysis embodies
9,000 kg CO_2e: Samaras and Meisterling. "Life Cycle Assessment
of Greenhouse Gas Emissions from Plug-in Hybrid Vehicles." Since
a new Prius costs about $24,000, its embodied emissions are about
0.4 kg CO_2e per dollar—roughly equivalent to the rule of thumb.

43. In a year, the average American spends $2,200, $1,600, and $600 on
home furnishings (including things such as towels and computers),
clothes, and personal care products, respectively: ValuePenguin.com.
"Average Household Budget in the US." [online]. valuepenguin.com
/average-household-budget. The same person spends $1,900 in one
year toward buying a new car (see the next note). The total is $6,300.

44. To estimate that, on average, people in the US spend $1,900 per year
on new cars:
a) 17.5 million cars and light trucks were sold in the US in 2015: Mike
Spector et al. "U.S. Car Sales Set Record in 2015." *Wall Street Journal*,
January 5, 2016. [online]. wsj.com/articles/u-s-car-sales-poised-for
-their-best-month-ever-1451999939.
b) The average price of a vehicle was about $34,000: Douglas A.
McIntyre. "GM Able to Raise Average Car Price by 8%." 24/7WallSt
website, February 3, 2015. [online]. 247wallst.com/autos/2015/02/03
/gm-able-to-raise-average-car-price-by-8/.
c) The US population was 320 million in 2015: Robert Schlesinger.
"The 2015 U.S. and World Populations." *US News and World Report*,
December 31, 2014. [online]. usnews.com/opinion/blogs/robert
-schlesinger/2014/12/31/us-population-2015–320-million-and-world
-population-72-billion.

45. Mike Berners-Lee. "What's the carbon footprint of…building a
house." *Guardian*, October 14, 2010. [online]. theguardian.com/envir
onment/green-living-blog/2010/oct/14/carbon-footprint-house.
Note that our rule of thumb would suggest a price of $160,000 for this

house, not including the price of the land, but our rule of thumb can't account for large variations in regional real estate markets.

46. The garbage you and I throw away is known as municipal solid waste (MSW). Interestingly, for every tonne of MSW thrown out in the US, there are 40 tonnes of waste created upstream. Ninety-four percent of US waste is industrial (discards from manufacturing, mining, metals processing, fossil fuel procurement, and agriculture), 3.5% is from construction and demolition, and 2.5% is MSW: Annie Leonard. *The Story of Stuff: The Impact of Overconsumption on the Planet, Our Communities, and Our Health—And How We Can Make It Better*. Free Press, 2011, p. 186.

47. In 2014, landfills in the US generated 6.6 billion kg of methane: US EPA. "Inventory of U.S. Greenhouse Gas Emissions and Sinks: 1990–2014." Table 7-2. [online]. epa.gov/sites/production/files/2016-04/documents/us-ghg-inventory-2016-chapter-7-waste.pdf. Multiplying by our mean GWP of 65 and dividing by 320 million Americans yields 1,300 kg CO_2e per person. Of our organic waste, 35% is food waste (500 kg CO_2e), 25% is wood pulp (300 kg CO_2e), 25% is textiles (300 kg CO_2e), and 15% is yard waste (200 kg CO_2e): Percentages calculated from ibid., Table 7-6.

48. World Resources Institute. CAIT Climate Data Explorer.

49. In 2000, producing the electricity to run US sewage treatment plants emitted 15.5 megatonnes of CO_2e: Center for Sustainable Systems. "U.S. Wastewater Treatment Factsheet." University of Michigan Pub No. CSS04-14, 2016. [online]. css.snre.umich.edu/sites/default/files/U.S._Wastewater_Treatment_Factsheet_CSS04-14.pdf. Public waste water treatment plants in the US emitted an additional 38 megatonnes CO_2e from methane: EPA. "Inventory of U.S. Greenhouse Gas Emissions and Sinks"; I've scaled this to a methane GWP of 65. About ⅔ of sewage treatment emissions are from domestic wastewater. The rest is from industrial sources, mainly meat production (76%) and paper manufacturing (17%): EPA. "Inventory of U.S. Greenhouse Gas Emissions and Sinks," Table 7-18.

50. Of the organic material that goes into US landfills, 35% is food, 25% is textiles, 25% is paper, and 15% is yard waste: EPA. "Inventory of U.S. Greenhouse Gas Emissions and Sinks." Assuming these materials produce methane at similar rates when they decompose, they produce 500 kg CO_2e, 300 kg CO_2e, 300 kg CO_2e, and 200 kg CO_2e, respectively. Composting the yard waste and one-third of the food waste (the post-consumer food waste) therefore prevents 400 kg CO_2e of methane emissions.

51. These estimates were derived from data in: EPA. "Inventory of U.S. Greenhouse Gas Emissions and Sinks." See previous note.

52. In 2014, Amtrak used 2,200 Btu of energy per passenger mile: Stacy C. Davis et al. *Transportation Energy Data Book*, ed. 35. Oak Ridge National Laboratory, Pub #ORNL-6992, 2016, Table 9.10. One gallon of diesel has 137,000 Btu: US EIA. "Energy Explained." [online]. eia.gov /energyexplained/index.cfm/index.cfm?page=about_energy_units. The combustion of this fuel creates 12.1 kg CO_2, which means in 2014 Amtrak emitted 0.19 kg CO_2 per passenger mile. Assuming that ⅔ of the passengers are in coach and that they take ½ the space of sleeper car passengers (approximately true for the Coast Starlight train) and are therefore responsible for ½ the emissions means Amtrak train travel emits 0.14 kg CO_2 per coach passenger mile and 0.28 kg CO_2 per sleeper passenger mile.

53. 184 passenger miles per gallon of diesel: Amy Zipkin, "Smoothing the rides on greyhound," *New York Times*, May 17, 2008. [online]. nytimes.com/2008/05/17/business/17interview-long.html.

54. Davis et al. *Transportation Energy Data Book*, Table 9.10.

55. David J.C. MacKay. *Sustainable Energy: Without the Hot Air*. UIT, 2008, the chapter entitled "Planes II."

56. EPA. "Inventory of U.S. Greenhouse Gas Emissions and Sinks." They use a methane GWP of 25; for consistency with this book, I've scaled their methane estimate to a GWP of 65 by multiplying it by a factor of 2.6.

57. Calculation values and factors are based on US data. If you rent, you might need to coordinate with your landlord to estimate electricity and natural gas usage.

58. Anders Nordelöf et al. "Environmental impacts of hybrid, plug-in hybrid, and battery electric vehicles: What can we learn from life cycle assessment?" *International Journal of Life Cycle Assessment* 19(11) (2014). [online]. doi:10.1007/s11367-014-0788-0. Note that gas cars are estimated to have five tonnes CO_2e of embodied emissions, but this estimate will vary greatly depending on the size of the vehicle.

59. Recall that the US per capita vehicle miles of travel in 2013 was 9,400 miles, and that the average passenger vehicle on the road in the US in 2013 gets 21.6 miles per gallon. Therefore, the average vehicle burns 700 gallons of gas after 1.6 years.

60. National Renewable Energy Laboratory. "PV FAQs: What is the energy payback for PV?" [online]. nrel.gov/docs/fy05osti/37322.pdf. With an expected lifetime of 30 years, though, this is still a good deal.

61. As I explained in Chapter 5, roughly one-half of global emissions

growth comes from growth in consumption. The other one-half comes from population growth.

62. For 2009, for children aged 5–15 yrs: Nancy McGuckin. "Travel to School in California: Findings from the California—National Household Travel Survey." Prepared for Active Living Research, Bikes Belong Foundation, and the Safe Routes to School National Partnership, 2013, p. 13. [online]. travelbehavior.us/Nancy-pdfs/Travel to School in California.pdf.

63. One parent I know drives 20 miles each way—five tonnes of CO_2 and thousands of dollars in fuel costs for getting her kids to school! Having a uniformly strong public education system, with good schools in *every* neighborhood, isn't just good for our kids and for the future of the US, it's also good for the climate.

64. The US average fuel consumption of model-year 2008 vehicles sold (as opposed to on the road) was 20.8 miles per gallon; in 2016, it was 25.1 mpg.: University of Michigan Transportation Research Institute. [online]. umich.edu/~umtriswt/EDI_sales-weighted-mpg.html. I have used a value of 24.7 mpg in my calculations. Changes in average fuel consumption are driven by gasoline prices, although there is likely a lag time between gasoline price shifts and sales of larger vs. smaller vehicles.

65. Alliance for Water Efficiency. "Showering to Savings." [online]. home-water-works.org/indoor-use/showers.

66. Estimated as follows:
    (1) I took a baseline measurement with just the pilot lights running. (I turned the hot water heater to "vacation" mode so it wouldn't switch on.) After 65 minutes, 1.7±0.05 cubic feet had elapsed, a rate of 1.6 cubic feet per hour. (The uncertainty comes from my time measurement; I don't know the uncertainty in the meter.) This amounts to 38±1 cubic feet per day, about 0.4 therms.
    (2) I turned off the heater pilot light. With the other two pilot lights running, the meter elapsed 0.85±0.05 cubic feet in one hour (77 therms per year). So the heater pilot light had been accounting for ½ of the instantaneous total, but we run this pilot less than ½ the year. Averaging the two always-on pilots, I estimate that a typical pilot emits 460 kg CO_2e per year.

67. Suppose we want to heat 30 gallons of water from 60°F (16°C) to 120°F (49°C). It takes 8.33 BTU of thermal energy to raise 1 gallon of water 1°F, so our load requires 15,000 BTU. Burning 1 therm of natural gas yields 100,000 BTU, so our load requires burning 0.15 therms of natural gas, which releases 2 kg CO_2e.

68. Thanks to Parke Wilde for these arguments.
69. This tree planting option is likely a best-case scenario. For example, a company called Terrapass has several projects it suggests it will spend your money on. One Terrapass project (which I picked at random from their website) aims to provide a small amount of habitat for migratory birds in California. While this sounds like a nice project in its own right, it doesn't remove CO_2 from the air. A secondary aim of this project is to provide "an important new revenue stream for rice growers." Well, setting aside the fact that we shouldn't be growing water-intensive rice in drought-stricken California in the first place, or the fact that conventional rice agriculture is a significant source of methane, could providing revenue to industrial agriculture corporations (with their large carbon footprints) really offset the flights of Terrapass's customers?
70. Hunt and McFarlane. "'Peak soil.'"
71. The Earth's biomass contains 450–650 GtC, mostly in trees. But by 2015, humans had emitted 560 GtC from burning fossil fuels. To deal with this by planting trees, we'd have to add an additional global forest approximately equal to all the trees alive today, in addition to the reforestation to the 1750 state. Where would we put all those extra trees?
72. Elisabeth Rosenthal. "Paying more for flights eases guilt, not emissions." *New York Times*, November 17, 2009. [online]. nytimes.com /2009/11/18/science/earth/18offset.html.
73. "Easily" here does not mean "trivially." During the four-year period in which I was developing my methods, it was like having a moderately intense hobby. After this initial period, it became easy to maintain.

Chapter 10: Slow Travel

1. In addition, I carried a sleeping bag and pad, an extra shirt, an extra pair of shorts, a stash of surplus powdered Gatorade and Power Bars from the summer job, and not much else. I started off with a rain fly for shelter, but after a few days of biking against headwinds, I realized it wasn't worth its weight, and I mailed it to Chicago.
2. The following book was helpful: Forest Gregg. *SVO: Powering Your Vehicle with Straight Vegetable Oil*. New Society, 2008.
3. Some "greasers" deal with the viscosity issue by blending other fuels, such as gasoline, into the veggie oil.
4. Maeby, she'll get there. Maybe she won't.
5. The viscosity of WVO can also be lowered through chemical means; biodiesel is essentially WVO that has been chemically modified for lower viscosity: Lyle Estill and Bob Armantrout. *Backyard Biodiesel: How to Brew Your Own Fuel*. New Society, 2015.

6. Global production of vegetable oil: Jim Lane, "Global 2016/17 vegetable oil production to hit record level: USDA," *Biofuels Digest*, September 25, 2016. [online]. biofuelsdigest.com/bdigest/2016/09/25 /global-201617-vegetable-oil-production-to-hit-record-level-usda/.

   Yields per acre: Keith Addison. "Oil Yields and Characteristics." Journey to Forever. [online]. journeytoforever.org/biodiesel_yield .html.

7. My yard has about 1/20th of an acre of growing space, and I'd grow the sunflowers organically, which likely reduces yield by 40%: M. Mazzoncini. "Sunflower under conventional and organic farming systems: Results from a long term experiment in Central Italy." *Aspects of Applied Biology* 79, (2006). [online]. orgprints.org/10203/.

8. Maeby typically gets about 25 miles per gallon, but with the cubies on the roof, this dropped to about 22 miles per gallon.

9. Imagine all the gasoline you've burned in your life, together in one place. Multiply by the number of cars.

10. A few years earlier, one bright blue Tuesday morning, I did see some of the buildings fall. I'd stepped out of the Wall Street subway and onto that strip of island to a snowfall of singed papers falling gently from a burning World Trade Center. Sailing up to the island after the cruise, I tried to pick out the building I'd been sitting in as I talked on the phone with my dad, telling him I was OK, then repeating "please don't let the second building fall, please don't let the second building fall" as the second building fell, a lifetime ago.

11. The ship typically burns about 1.5 barrels per nautical mile. At the time of my trip, fuel was over $100 a barrel; fueling that round trip cost about $1,000,000.

12. According to Climate Care, a carbon offsetting company: "Is cruising any greener than flying?" *Guardian*, December 20, 2006. [online]. theguardian.com/travel/2006/dec/20/cruises.green. This number for fossil-fueled cruising, 0.4 kg CO_2 per passenger mile, is reasonable: it's what *Spirit* would emit if she carried 2,000 passengers (see next note).

13. The *Spirit* burns 1.5 barrels (63 gallons) of bunker fuel per nautical mile. Bunker fuel (residual fuel) emits 11.8 kg CO_2/gallon: US EIA. "Carbon Dioxide Emissions Coefficients." Release date, February 2, 2016. [online]. eia.gov/environment/emissions/co2_vol_mass.cfm. Adding 20% extra for upstream emissions (see Chapter 9), this comes to 14.1 kg CO_2 / gallon. So over one nautical mile, *Spirit* emits 892 kg CO_2, or 780 kg CO_2 per statute mile. She has a 29,500 short ton cargo capacity, so with my 50 pounds of baggage, I accounted for 0.0000041 of the cargo, making my emissions 0.0032 kg CO_2 per mile.

Chapter 11: Meditation, a Foundation of Change

1. In the vipassana tradition, there's no charge for courses, which run on a "pay it forward" basis through donations from people who have taken at least one course. The teachers and the people who run the course and cook the food are volunteers. This commitment to non-commercialism and selflessly helping others maintains the purity of the teaching. For more information: dhamma.org.

2. S. N. Goenka. "The Art of Living: Vipassana Meditation." [online]. dhamma.org/en-US/about/art.

3. Robert A. F. Thurman. *Essential Tibetan Buddhism*, rev. ed. Harper-One, 1996.

4. Taking intoxicants may seem relatively harmless. However, being intoxicated makes it easier to break the other four precepts. For example, drinking makes it much easier to act violently. Also, I have friends who've received drunk driving convictions, which caused them great suffering.

5. For me, the transition away from breaking these precepts was gradual. For example, early in my meditation practice, I drank alcohol every day. As I practiced, I became aware that alcohol gave me a feeling of mental cloudiness and sapped my energy. I preferred feeling sober, so I gradually started drinking less. I've become sensitive to the distinct taste and smell of alcohol, and I don't like it. If you love beer and wine, as I did, you don't have to be afraid to give them up. But when you do drink, you will probably have a heightened awareness of the associated sensations, and you might choose to drink less.

6. There are studies based on other meditative techniques which also demonstrate benefits. There are few head-to-head studies comparing different meditation techniques.

7. Britta K. Hölzel et al. "Mindfulness practice leads to increases in regional brain gray matter density." *Psychiatry Research: Neuroimaging* 191 (2011). [online]. doi:10.1016/j.pscychresns.2010.08.006; Sarah W. Lazar et al. "Meditation experience is associated with increased cortical thickness." *Neuroreport* 16 (2005). [online]. ncbi.nlm.nih.gov/pmc/articles/PMC1361002/.

8. Hölzel et al. "Mindfulness practice leads to increases in regional brain gray matter density."

9. Yvette I. Sheline et al. "Depression Duration But Not Age Predicts Hippocampal Volume Loss in Medically Healthy Women with Recurrent Major Depression." *The Journal of Neuroscience* 19 (1999). [online]. jneurosci.org/content/19/12/5034.full.

10. For example: Zindel V. Segal et al. *Mindfulness-Based Cognitive Therapy for Depression*, 2nd ed. Guilford Press, 2012.

11. While in my early twenties, I had a bout of severe depression. I cannot begin to express the awfulness of that experience; I'm grateful to have made it out alive. Twice since I began meditating in 2003, during periods of stress during which I wasn't keeping my regular meditation, I felt a sort of mental pain or pressure that reminded me of being depressed. It felt as though depression were trying to find its way back in. Both times, simply reinstating my meditation routine caused that feeling to vanish after a day or two. What's more, my meditation experience is what allowed me to quickly recognize the incipient mental sensation in the first place.

12. Hölzel et al. "Mindfulness practice leads to increases in regional brain gray matter density."

13. Philippe R. Goldin and James J. Gross. "Effects of Mindfulness-Based Stress Reduction (MBSR) on Emotion Regulation in Social Anxiety Disorder." *Emotion* 10(1) (2010). [online]. doi:10.1037/a0018441.

14. T. L. Jacobs et al. "Self-reported mindfulness and cortisol during a Shamatha meditation retreat." *Health Psychology* 32 (10) (2013). [online]. doi:10.1037/a0031362.

15. Lazar et al. "Meditation experience is associated with increased cortical thickness."

16. Ibid.

17. T. Gard et al. "The potential effects of meditation on age-related cognitive decline: A systematic review." *Annals of the New York Academy of Sciences* 1307 (2014). [online]. doi:10.1111/nyas.12348.

18. Judson A. Brewer et al. "Meditation experience is associated with differences in default mode network activity and connectivity." *Proceedings of the National Academy of Sciences* 108 (2011). [online]. doi:10.1073/pnas.1112029108.

19. Matthew A. Killingsworth and Daniel T. Gilbert. "A wandering mind is an unhappy mind." *Science* 330 (2010). [online]. doi:10.1126/science.1192439.

20. Brewer et al. "Meditation experience is associated with differences in default mode network activity and connectivity."

21. N. E. Morone et al. "Mindfulness meditation for the treatment of chronic low back pain in older adults: A randomized controlled pilot study." *Pain* 134(3) (2008). [online]. doi:10.1016/j.pain.2007.04.038.

22. David S. Black et al. "Mindfulness Meditation and Improvement in Sleep Quality and Daytime Impairment Among Older Adults With Sleep Disturbances: A Randomized Clinical Trial." *Journal of the American Medical Association Internal Medicine* 175 (2015). [online]. doi:10.1001/jamainternmed.2014.8081. Also, vipassana cured my mom's persistent insomnia.

23. Richard J. Davidson et al. "Alterations in Brain and Immune Function Produced by Mindfulness Meditation." *Psychosomatic Medicine* 65(4) (2003). [online]. doi:10.1097/01.PSY.0000077505.67574.E3.

24. Joel W. Hughes et al. "Randomized Controlled Trial of Mindfulness-Based Stress Reduction for Prehypertension." *Psychosomatic Medicine* 75(8) (2013). [online]. doi:10.1097/PSY.0b013e3182a3e4e5.

Chapter 12: Reconnecting with Mother Earth

1. Carolyn Dimitri et al. *The 20th Century Transformation of U.S. Agriculture and Farm Policy.* USDA Economic Information Bulletin Number 3, June 2005. [online]. ilovefarmers.org/downloads/The20th CenturyTransformationofU.S.AgricultureandFarmPolicy.pdf.

2. It's one thing to grow fruits and veggies in a tiny yard, but grains had me perplexed. I failed at growing Sonora wheat, an heirloom well-suited to Southern California's hot and dry conditions. Meantime, a sorghum volunteer, whose seed must have been pooped out by a bird, began to grow. I had no idea what it was, but it grew up on its own. I have a policy of not pulling out plants until I know what they are, and it paid off this time: eventually it dawned on me that I'd found an ideal grain crop. Or rather, it had found me.

3. A good guide: Eric Toensmeier. *Perennial Vegetables from Artichokes to Zuiki Taro: A Gardener's Guide to Over 100 Delicious, Easy-to-Grow Edibles.* Chelsea Green, 2007.

4. Masanobu Fukuoka. *The One-Straw Revolution: An Introduction to Natural Farming.* Rodale Press, 1978.

5. Ibid., p. 52.

6. Ibid., p. 109.

7. Roc Martin. "The Amish Farmers Reinventing Organic Agriculture." *Atlantic*, October 6. 2014. [online]. theatlantic.com/health/archive /2014/10/the-amish-farmer-replacing-pesticides-with-nutrition /380825//.

8. Humanure composting might break a municipal code where you live. Proceed at your own risk.

9. In one week, I use 3 or 4 gallons of water rinsing buckets. Back in my flush toilet days, I'd use over 100 gallons in a week (30 or so flushes at around 4 gallons per flush). What's more, those rinse water gallons aren't wasted: they support microbes working day and night to build soil.

10. Joseph C. Jenkins, *The Humanure Handbook: A Guide to Composting Human Manure, 3rd ed.* Jenkins Publishing, 2005. Jenkins, in turn, is indebted to an obscure World Bank treatise which lays out the basics of human manure composting with key data on hygiene: Richard G.

Feachem et al. Transportation, Water and Telecommunications Department. *Appropriate technology for water supply and sanitation: Health aspects of excreta and sullage management: A state-of-the-art review.* World Bank, 1981. [online]. documents.worldbank.org/curated /en/929641467989573003/Appropriate-technology-for-water-supply -and-sanitation-health-aspects-of-excreta-and-sullage-management-a -state-of-the-art-review.

11. Many restaurants throw out five-gallon buckets; you can easily get as many as you want without using money. The box can be made from scrap or salvaged wood.

12. World Health Organization. "Cholera Fact Sheet" Updated October 2016. [online]. who.int/mediacentre/factsheets/fs107/en/.

13. Feachem, "Appropriate Technology," p. 105.

14. During this experiment, nightly lows in Altadena were around 50°F and daily highs were around 70°F.

15. Humanure composting temperatures should be high enough to kill most weed seeds. In a study of six weed seeds, even the hardiest experienced 100% mortality after only three hours at 140°F (60°C): Ruth M. Dahlquist et al. "Time and Temperature Requirements for Weed Seed Thermal Death." *Weed Science* 55(6) (2007). [online]. doi:10.1614/WS-04-178.1.

16. After two weeks, it was 130°F. After a month, it was around 110°F. I poured some water near the thermometer, and it went back to 130°F.

17. World Health Organization. "Sanitation Fact Sheet" Reviewed November 2016. [online]. who.int/mediacentre/factsheets/fs392/en/.

18. Jay P. Graham and Matthew L. Polizzotto. "Pit latrines and their impacts on groundwater quality: A systematic review." *Environmental Health Perspectives* 121(5) (2013). [online]. doi:10.1289/ehp.1206028. This risk, although serious, remains poorly quantified.

19. World Health Organization. "Sanitation Fact Sheet."

20. Nayantara Narayanan. "Horrifying fact: Almost all India's water is contaminated by sewage."*Scroll.in,* July 1, 2015. [online]. scroll.in /article/737981/horrifying-fact-almost-all-indias-water-is-contami nated-by-sewage/.

21. US EPA. "Septic Systems Overview." [online]. epa.gov/septic/septic -systems-overview.

22. Marylynn V. Yates. "Septic Tank Density and Ground-Water Contamination." *Groundwater* 23(5) (1985). doi:10.1111/j.1745-6584.1985 .tb01506.x.

23. Chenxi Wu et al. "Uptake of Pharmaceutical and Personal Care Products by Soybean Plants from Soils Applied with Biosolids and Irrigated with Contaminated Water." *Environmental Science &*

Technology 44 (2010). [online]. doi:10.1021/es1011115. Pharmaceuticals in this context are still poorly understood, part of a broad class of "emerging contaminants to the environment." (Quote from USGS. "Land Application of Municipal Biosolids." [online]. toxics.usgs.gov /regional/emc/municipal_biosolids.html.)

24. Not all cultures have shared our taboo about poop. F. H. King (in *Farmers of Forty Centuries: Organic Farming in China, Korea, and Japan*. 1911, repr., Dover, 2004) recounted that in ancient China night soil collectors would actually pay for the privilege of emptying privies, I assume from folks without gardens. Until recently, Mandarin Chinese usage did not include profane words for *poop*, as Western languages do. This uncomposted night soil wasn't hygienic, but it did allow the ancient Chinese to develop a sustainable agriculture—something industrial civilization has so far failed to do.

25. My chickens love the larvae of the figeater beatle. My late favorite hen, Black Star, could jump and grab flying adults out of the air when they flew too low.

26. See David Quammen. *Spillover: Animal Infections and the Next Human Pandemic*. Norton, 2012.

27. Maryn McKenna. "The looming threat of avian flu." *New York Times Magazine*, April 13, 2016. [online]. nytimes.com/2016/04/17 /magazine/the-looming-threat-of-avian-flu.html.

28. Mary J. Gilchrist et al. "The potential role of concentrated animal feeding operations in infectious disease epidemics and antibiotic resistance." *Environmental Health Perspectives* 115(2) (2007). [online]. doi:10.1289/ehp.8837.

29. An 18-inch box with a wire screen taped over a hole for ventilation.

30. Thomas D. Seeley. *Honeybee Democracy*. Princeton, 2010.

31. About half of my swarm captures stay; the other half will fly off after a day or two, probably to find a home more to their liking. I'll arrive to find an empty box. I've recently started to use a piece of queen excluder over the entrance to keep the queen inside for the first few days, but I haven't done this enough to say for sure if it helps.

32. Michael Bush. "Genetic Diversity in Bees." The Practical Beekeeper website, 2008. [online]. bushfarms.com/beesgeneticdiversity.htm.

33. See Chapter 5.

34. Dennis vanEngelsdorp et al. "Colony Collapse Disorder: A Descriptive Study." *PLoS ONE* 4(8) (2009). [online]. doi:10.1371/journal .pone.0006481.

35. I used to keep two hives on my roof as well. If you decide to go the roof route, make sure the hives have shade, and realize that a mature

hive full of honey could weigh 200 pounds and present an interesting challenge to move.

Chapter 13: Opting Out of a Broken System

1. Around the same time, I added extra insulation to our attic floor, which reduces transmission heat from the hot attic to the living space, and makes sense in any climate. Other energy-saving cooling alternatives (that may work best in hot, dry climates) include a whole-house fan, radiant barrier under attic rafters, and a reflective "cool roof," which has the additional benefit of reducing the urban heat island effect. If I ever replace my roof, it will be with a cool roof: heat waves are a bigger challenge in Southern California than cold snaps.

2. Access to machine shops is a perk of being part of a university community.

3. For example, see Mrs. Homegrown (a.k.a. Kelly Coyne). "A Homemade Mattress." Root Simple blog, March 15, 2013. [online]. root simple.com/2013/03/a-homemade-mattress/.

4. See Jon Jandai. "Life is easy. Why do we make it so hard?" TEDxDoi Suthep, August 3, 2011. [online]. youtube.com/watch?v=21j_OCNLuYg.

5. Edited for clarity from Richard Whittaker. "A Conversation with Adam Campbell: A Taste For Life." conversations.org website, November 30, 2012. [online]. conversations.org/story.php?sid=336.

6. See for example: Keiren. "Rocket Mass Heaters." Blog entry. [online]. niftyhomestead.com/blog/rocket-mass-heaters/.

7. Jon Jandai. "Life is easy." Also, Ianto Evans and Linda Smiley built a cob home for $500: Michael Smith and Ianto Evans. "Questions and Answers about Cob." Natural Building Colloquium Southwest. [online]. networkearth.org/naturalbuilding/cob.html.

8. Stores overstock before holidays and then throw out two days' worth of expiring food instead of one.

9. Organic doesn't necessarily mean that pesticides weren't used. Still, I'd rather eat an organic apple than a conventional one. See: Beth Hoffman. "Five reasons to eat organic apples: Pesticides, healthy communities, and you." *Forbes*, April 23, 2012. [online]. forbes.com /sites/bethhoffman/2012/04/23/five-reasons-to-eat-organic-apples -pesticides-healthy-communities-and-you/#5b19cc846d21. As far as food certifications go, organic is the best we have; we should work together to continue strengthening it.

10. This takes an hour of my time every other week. I got this gig after I offered bread to my crop-swap group after a Christmas dumpster dive. Someone in the group was already working with the

supermarket and asked me to help. This is how the universe works: when you are open to people, people are open to you.

11. Dana Gunders. "Wasted: How America Is Losing up to 40 Percent of Its Food from Farm to Fork to Landfill." Natural Resources Defense Council Issue Paper #IP-12-06-B, August 2012. [online]. nrdc.org /sites/default/files/wasted-food-IP.pdf.

12. As far as preserving food goes, I also like canning homegrown tomatoes, drying fruit such as figs (laying trays of fruit on a car dashboard keeps the flies off), and saving up vegetable scraps in freezer bags to make broth.

13. There's some evidence that safrole is carcinogenic in rats, at ingestion levels far greater than what you'd get from homemade root beer: Peter G. Wislocki et al. "Carcinogenic and Mutagenic Activities of Safrole, 1'-Hydroxysafrole, and Some Known or Possible Metabolites." *Cancer Research* 37(6) (1977). [online]. cancerres.aacrjournals .org/content/37/6/1883.short. Many people use sassafras anyway.

14. I then obtained a card through the Permaculture Credit Union, the only card I could find with no connection to a large bank. That credit union is now managed by the Sandia Area Credit Union.

15. Sharon was inspired by the 40-day period of fasting observed during Christian Lent.

16. Luke Hurst. "HSBC warns clients of fossil fuel investment risks." *Newsweek*, April 21, 2015. [online]. newsweek.com/hsbc-warns-clients -fossil-fuel-investment-risks-323886.

17. War Resisters League. "U.S. Federal Budget—2015 Fiscal Year: Where Your Income Tax Money Really Goes." [online]. warresisters.org /sites/default/files/2015 pie chart—high res.pdf.

18. The W2 system makes war tax resistance more difficult. Take too many exemptions, and you risk being charged with W2 fraud.

19. New York Yearly Meeting. "Purchase Quarterly Meeting of the Religious Society of Friends (Quakers): Peace Tax Escrow Account." [online]. nyym.org/purchasequarter/peacetax.html/.

20. Mark Koba. "U.S. military spending dwarfs rest of world." *NBC News*, February 24, 2014. [online]. nbcnews.com/storyline/military -spending-cuts/u-s-military-spending-dwarfs-rest-world-n37461.

21. Meredith Bennett-Smith. "Womp! This Country Was Named the Greatest Threat To World Peace." *Huffington Post*, January 2, 2014. [online]. huffingtonpost.com/2014/01/02/greatest-threat-world -peace-country_n_4531824.html.

22. Daniel Suelo interview by Shirley Jahad. Crawford Family Forum, Pasadena, California, March 26, 2012. Archived by Southern Cali-

fornia Public Radio. [online]. scpr.org/events/2012/03/26/584/one -man-quit-money/.

23. The total time spent staring at screens in the US is closer to eight hours: Molly Brown. "Nielsen reports that the average American adult spends 11 hours per day on gadgets." GeekWire, March 13, 2015. [online]. geekwire.com/2015/nielsen-reports-that-the-average -american-adult-spends-11-hours-per-day-on-gadgets/. The twelve years statistic assumes eight hours of sleep per day.

24. "Advertising to Children and Teens: Current Practices: A Common Sense Media Research Brief." Common Sense Media, January 28, 2014. [online]. commonsensemedia.org/research/advertising-to -children-and-teens-current-practices.

25. Bruce E. Levine, "Does TV actually brainwash Americans?" *Salon*, October 30, 2012. [online]. salon.com/2012/10/30/does_tv_actually _brainwash_americans/.

26. Mengwei Bian and Louis Leung. "Linking Loneliness, Shyness, Smartphone Addiction Symptoms, and Patterns of Smartphone Use to Social Capital." *Social Science Computer Review* 33(1) (2015). [online]. doi:10.1177/0894439314528779.

27. Kimberly S. Young and Robert C. Rogers. "The Relationship Between Depression and Internet Addiction." *CyberPsychology and Behavior*. 1(1) (1998). [online]. doi:10.1089/cpb.1998.1.25.

28. If you do see such an action, please let me know what it is.

Chapter 14: Collective Action

1. Tom Mintier and Reuters. "Global warming pact approved, develop- ing nations face few restrictions." *CNN*, December 11, 1997. [online]. cnn.com/EARTH/9712/11/climate.conf/index.html.

2. Carbon Dioxide Information Analysis Center. "Global Fossil-Fuel CO_2 Emissions." [online]. cdiac.ornl.gov/trends/emis/tre_glob_2013 .html.

3. Joeri Rogelj et al. "Paris Agreement climate proposals need a boost to keep warming well below 2°C." *Nature* 534 (2016). [online]. doi:10.1038/nature18307.

4. See IGM Economic Expert Panel. "Carbon Tax." IGM Chicago, De- cember 20, 2011. [online]. igmchicago.org/surveys/carbon-tax; Luca Taschini et al. "Carbon tax v cap-and-trade: Which is better?" *Guard- ian*, January 31, 2013. [online]. theguardian.com/environment/2013 /jan/31/carbon-tax-cap-and-trade.

5. Calculated this way: $30 per tonne CO_2 times 0.0088 tonnes CO_2 per gallon, with an extra 28% for upstream emissions (see Chapter 9).

6. Calculated this way: $30 per tonne CO_2 times 20 tonnes CO_2e per person in the US.

7. Depending on the specifics of the policy, of course, children might not necessarily receive a full share. For example, at least one proposal calls for a half share for the first two children; in this case, a household with two adults and two children would receive $9,000 from a carbon price of $150 per tonne CO_2e.

8. Carbon Tax Center. "Ensuring Equity." [online]. carbontax.org/?s =ensuring+equity.

9. Carbon Tax Center. "Dividends." [online]. carbontax.org/?s =dividends.

10. A catalog with these carbon estimates would need to be assembled. In addition to making the border adjustment possible, by creating this catalog, we would learn a great deal about how carbon is em- bodied in our economy. The border adjustment would initially target carbon-intensive goods: iron, steel, cement, glass, paper, etc. It's easy to calculate the adjustment in these cases, and together they capture the majority of embodied emissions. Estimates for goods with lower carbon intensities can be continually refined. This border adjustment would be consistent with WTO rules (i.e., "most favored nation" and "national treatment"). Also, the border adjustment money would stay separate from the fee and dividend money. The tariffs collected from imported goods would be used to pay refunds to domestic producers exporting to "dirty regimes." In the case of the US, given the US trade deficit, this pot of money would likely grow over time; it would be up to Congress to decide how to spend the windfall. Exports of domes- tically produced fossil fuels themselves could be exempt from the border adjustment refund, putting upward pressure on the price of fossil fuels and further discouraging their use, even internationally.

11. R. A. "Do economists all favour a carbon tax?" *Economist*, Septem- ber 19, 2011. [online]. economist.com/blogs/freeexchange/2011/09 /climate-policy.

12. Goldman Sachs has lobbied hard for cap-and-trade. That should tell you something: Matt Taibbi. "The Great American Bubble Machine." *Rolling Stone*, April 5, 2010. [online]. rollingstone.com/politics/news /the-great-american-bubble-machine-20100405.

13. Although emissions in Europe went down, the EU ETS probably wasn't a significant cause: Olivier Gloaguen and Emilie Alberola. "As- sessing the factors behind CO_2 emissions changes over the phases 1 and 2 of the EU ETS: An econometric analysis." CDC Climat Recher- che Working Paper No, 2013–15, October 2013. [online]. cdcclimat

.com/IMG/pdf/13-10_cdc_climat_r_wp_13-15_assessing_the
_factors_behing_co2_emissions_changes.pdf.

14. E.g., Paul Krugman. "Unhelpful Hansen." *New York Times*, December 7, 2009. [online]. krugman.blogs.nytimes.com/2009/12/07/un helpful-hansen.

15. P. F. "British Columbia's carbon tax: The evidence mounts." *Economist*, July 31, 2014. [online]. economist.com/blogs/americasview /2014/07/british-columbias-carbon-tax.

16. Dr. Stewart Elgie and Jessica McClay. "Policy Commentary/Commentaire BC's Carbon Tax Shift Is Working Well after Four Years (Attention Ottawa)." *Canadian Public Policy 39(2)* (2013). [online]. energyindependentvt.org/wp-content/uploads/2014/11/BC_Car bon-Tax-success-story.pdf.

17. Scott Nystrom and Patrick Luckow. "The Economic, Climate, Fiscal, Power, and Demographic Impact of a National Fee-and-Dividend Carbon Tax." Regional Economic Models, Inc. (REMI) and Synapse Energy Economics, Inc. [online]. citizensclimatelobby.org/wp-con tent/uploads/2014/06/REMI-carbon-tax-report-62141.pdf. Note that this study was funded by Citizens' Climate Lobby, an advocacy group, but that REMI has a reputation for being unbiased.

18. Ibid.; and Marc Breslow et al. "Analysis of a Carbon Fee or Tax as a Mechanism to Reduce GHG Emissions in Massachusetts." prepared for the Massachussetts Department of Energy Resources, December, 2014. [online]. mass.gov/eea/docs/doer/fuels/mass-carbon-tax-study .pdf.

19. Statistics Canada, via P. F. "British Columbia's carbon tax." Note that the BC carbon fee is implemented as a sales tax, and returns 100% of revenues by lowering taxes.

20. Greg Mankiw. "How Not to Pass a Carbon Tax." Blog post, August 3, 2015. [online]. gregmankiw.blogspot.ca/2015/08/how-not-to-pass -carbon-tax.html.

21. Jerry Taylor. "The Conservative Case for a Carbon Tax." Niskanen Center, March 23, 2015. [online]. niskanencenter.org/wp-content /uploads/2015/03/The-Conservative-Case-for-a-Carbon-Tax1.pdf.

22. Ted Deutch and Carlos Curbelo. "Creating a bipartisan climate to discuss climate change in Congress." *The Hill*, March 24, 2016. [online]. thehill.com/blogs/congress-blog/energy-environment/274061 -creating-a-bipartisan-climate-to-discuss-climate.

23. Editorial Board. "Even Big Oil Wants a Carbon Tax." *Bloomberg View*, June 1, 2015. [online]. bloomberg.com/view/articles/2015-06-01/even -big-oil-wants-a-carbon-tax.

24. Helge Lund et al. "Letter to Ms. Christiana Figueres and Mr. Laurent Fabius." May 29, 2015. [online]. bp.com/content/dam/bp/pdf/press /paying-for-carbon.pdf.

25. Alan S. Blinder. "The Carbon Tax Miracle Cure." *Wall Street Journal*, January 31, 2011. [online]. wsj.com/articles/SB10001424052748703893 10457610861068157691 4.

26. Annually, 3.7 million deaths are caused by outdoor air pollution, which is mainly from burning fossil fuels: World Health Organization. "7 million premature deaths annually linked to air pollution." March 25, 2014. [online]. who.int/mediacentre/news/releases/2014 /air-pollution/en/. Annually, indoor and outdoor pollution together cost $5 trillion, not even counting healthcare costs: John Vidal. "Air pollution costs trillions and holds back poor countries, says World Bank." *The Guardian*, September 8, 2016. [online]. theguardian.com /global-development/2016/sep/08/air-pollution-costs-trillions-holds -back-poor-countries-world-bank.

27. Fabio Caiazzo et al. "Air pollution and early deaths in the United States. Part I: Quantifying the impact of major sectors in 2005." *Atmospheric Environment* 79 (2013). [online]. doi:10.1016/j.atmosenv .2013.05.081.

28. It only takes a few minutes to write a letter, and it's a lot of fun if it gets published. One of my letters, which pretty well sums up my perspective on climate action, can be found at: "Climate solutions: From marches to policies." *New York Times*, September 22, 2014. [online]. nytimes.com/2014/09/23/opinion/climate-solutions-from-marches -to-policies.html.

29. US EIA. "Frequently asked questions: How much of U.S. carbon dioxide emissions are associated with electricity generation?" [online]. eia.gov/tools/faqs/faq.cfm?id=77&t=11.

30. A terawatt (TW) is one trillion watts of power; a TWh is the energy needed to provide a TW of power for an hour. 1 TWh = 1 billion kWh. The 4,070 TWh figure is from: US EIA. "Electric Power Annual 2013." March 2015. [online]. eia.gov/electricity/annual/archive/03482013 .pdf.

31. Ibid.: Nuclear: 790 TWh; hydroelectric: 270 TWh; non-hydroelectric renewables: 250 TWh. Wind produced 168 TWh, or 4.1% of the total, while solar power (thermal and photovoltaic) produced 9 TWh, or 0.2% of the total.

32. Three-quarters of US electricity is consumed by residential and commercial users (with each sector using about the same amount) and ¼ is consumed by industry (US EIA. "Annual Energy Outlook 2017."

[online]. eia.gov/outlooks/aeo/). In our household, we use less than ¹⁄₁₀ the electricity per person than the US average. While this argues for easy and deep reductions in the residential sector, I suspect the commercial sector could easily make similar reductions: the lights in many office buildings stay on all night, for example.

33. In 2013, the US used 6,900 kg oil equivalent per person per year, whereas the UK used 3,000: World Bank. "Energy use (kg of oil equivalent per capita)." [online]. data.worldbank.org/indicator/ EG.USE.PCAP.KG.OE.

34. Calculated this way: $(4{,}070 \text{ TWh/yr} / 2) - (790 \text{ TWh/yr} + 270 \text{ TWh/yr} + 250 \text{ TWh/yr})$

35. Oil Change International. "Fossil Fuel Subsidies: Overview." [online]. priceofoil.org/fossil-fuel-subsidies/.

36. David Coady et al. "How Large Are Global Energy Subsidies?" International Monetary Fund working paper #WP/15/105. [online]. imf .org/external/pubs/ft/wp/2015/wp15105.pdf.

37. Our Children's Trust. [online]. ourchildrenstrust.org/. The suit makes two claims. First, a constitutional claim that "the government's aggregate actions, including subsidizing, permitting and facilitating the development, transportation, and combustion of fossil fuels, violates the youths' substantive due process rights under the 5th Amendment's protection against deprivation of life, liberty and property without due process of law." The case isn't challenging the government's *inaction*; it challenges its direct *action* toward making global warming worse. Second, a public trust doctrine claim that "same actions...violate the federal government's fiduciary obligation to preserve essential natural resources for the benefit of all present and future generations."

38. Tony Dokoupil. "Big Oil joins legal fight against little kids over climate change." *MSNBC*, August 14, 2015, updated November 13, 2015. [online]. msnbc.com/msnbc/big-oil-joins-legal-fight-little-kids-over -climate-change.

39. Two charts were part of an exhibit the plaintiffs' attorneys presented to the court on September 13, 2016: Our Children's Trust. "Details of Proceedings." [online]. ourchildrenstrust.org/federal-proceedings/.

40. Sophia V. Schweitzer. "Are countries legally required to protect their citizens from climate change?" *Ensia*, July 15, 2015. [online]. ensia.com/features/are-countries-legally-required-to-protect-their -citizens-from-climate-change/.

41. Oxfam. "62 people own the same as half the world, reveals Oxfam Davos report." Press release, January 18, 2016. [online]. oxfam.org

/en/pressroom/pressreleases/2016-01-18/62-people-own-same-half
-world-reveals-oxfam-davos-report.

42. A billionaire would respond that the source of his wealth was his
brilliant idea. But doesn't that idea just provide the means with
which to extract (and therefore concentrate and accumulate) the
distributed wealth of labor or resources? Ideas can serve to mobilize
and rearrange labor and resources, but the idea by itself—without
that labor and those resources—does not constitute wealth. Thomas
Frank wrote: "Many of our most vaunted innovations are simply
methods—electronic or otherwise—of pulling off some age-old
profit-maximizing maneuver by new and unregulated means."
Thomas Frank. *Listen, Liberal: Or, What Ever Happened to the Party of
the People?* Metropolitan Books, 2016, p. 209.

43. Paul Davidson. "Decline of unions has hurt all workers: Study." *USA
Today*, August 30, 2016. [online]. usatoday.com/story/money/2016
/08/30/decline-unions-has-hurt-all-workers-study/89557266/.
Unions have done a world of good since their formation in mid-19th-
century America, bringing us such wonderful things as weekends, to
name but one example. Still, unions sometimes push their negoti-
ations too far, perhaps doing themselves more harm than good by
fueling an opposition that paints them as unreasonably greedy and
corrupt.

44. David Rotman. "How Technology Is Destroying Jobs." *MIT Technology Review*, June 12, 2013. [online]. technologyreview.com/s/515926
/how-technology-is-destroying-jobs/.

45. For most of US history, lobbying was strongly censured by the
courts, and has at times been illegal in some states: Alex Mayyasi.
"When Lobbying Was Illegal." *Priceonomics*, April 15, 2016. [online].
priceonomics.com/when-lobbying-was-illegal/.

46. For an excellent discussion of the problems arising from international
trade as we've structured it, see Annie Leonard. *The Story of Stuff.*

47. Meadows. "Leverage Points: Places to Intervene in a System."

48. Gilda Sedgh et al. "Intended and Unintended Pregnancies Worldwide
in 2012 and Recent Trends." *Studies in Family Planning* 45(3) (2014).
[online]. doi:10.1111/j.1728-4465.2014.00393.x.

49. There are deep connections between Christianity, Islam, and population growth. Muslims and Christians have the highest fertility rates of
any religion, at 3.1 and 2.7 children per woman, respectively (compared to the global average of 2.5 and the replacement rate of 2.1).
The world's third-largest religion, Hinduism, has a fertility rate of 2.4:
Pew Research Center. "The Future of World Religions: Population

Growth Projections, 2010—2050." April 2, 2015. [online]. pewforum
.org/2015/04/02/religious-projections-2010-2050/.

50. Wilson. *Half-Earth.*

51. Robert H. MacArthur and Edward O. Wilson. *Theory of Island Bioge-
ography.* Monographs in Population Biology 1, Princeton, 1967.

52. Thanks to Noam Chomsky for articulating this point: Noam
Chomsky. "Noam Chomsky on Trump and the decline of the American Superpower." YouTube, December 5, 2016. [online]. youtube.com
/watch?v=Yp74MQBGMnk.

Chapter 15: Community

1. Henry David Thoreau. *Walden.* Internet Bookmobile, 2004, Chapter 1, "Economy," pp. 25 and 26. [online]. eldritchpress.org/walden5
.pdf.

2. Meadows. "Leverage Points: Places to Intervene in a System."

3. The concept of "online community" seemed promising at first, and in
2009 or thereabouts, I started a "350 Los Angeles" Facebook group.
350.org is an organization of climate activists that takes its name from
the CO_2 atmospheric concentration deemed to be the maximum safe
level by climate scientist James Hansen (350 parts per million). While
Facebook activism might work well for some people, my own experience of "350 Los Angeles" on Facebook turned out to be pointless and
disempowering: I'd post diatribes that disappeared into the internet
void and organize events to which no one showed up.

4. A PVC pipe receives my washing machine drain hose; the hose
empties into a long narrow trench between my avocado and orange
trees that's filled with wood chip mulch (used only with sodium-free
detergents).

5. These circles were pioneered by my friend Nipun Mehta and have
grown into an organic network of gift economy: Awakin. [online].
awakin.org.

6. For links to groups, see: Transition Network. [online]. transition
network.org.

7. For details about projects in California, see: Clean Power Exchange.
[online]. cleanpowerexchange.org/.

8. Sonoma Clean Power [online]. sonomacleanpower.org/; Marin
Clean Energy [online]. mcecleanenergy.org/.

9. See the incredible short film: Ben C. Solomon and Tommy Tren-
chard. "Erison and the ebola soccer survivors," *New York Times,* 2015.
[online]. nytimes.com/video/world/africa/100000003815213/erison
-and-the-ebola-soccer-survivors.html.

10. Lara P. Clark et al. "National Patterns in Environmental Injustice and Inequality: Outdoor NO_2 Air Pollution in the United States." *PLoS ONE* 9(4) (2014). doi:10.1371/journal.pone.0094431.

11. For example: Heather Clancy. "TNC's Mark Tercek: Protect, transform and inspire." *GreenBiz*, July 9, 2015. [online]. greenbiz.com /article/tncs-mark-tercek-protect-transform-and-inspire.

12. Tami Luhby. "The black-white economic divide in 5 charts." *CNN Money*, November 25, 2015. [online]. money.cnn.com/2015/11/24 /news/economy/blacks-whites-inequality/index.html.

13. I am grateful to Daniel Suelo for teaching me about grace.

Index

About the Author

PETER KALMUS is a climate scientist working at NASA's Jet Propulsion Laboratory. He has a PhD in physics from Columbia University and a BA in physics from Harvard. At work, he uses satellite data, *in situ* data, and models to study the rapidly changing Earth, with a focus on boundary layer clouds. At home, he explores how dramatically reducing carbon emissions can lead to a happier, more connected life. He enjoys orcharding, beekeeping, and backpacking with his family.

About the Illustrator

SAM BOWER UI is a full-time volunteer trying to live simply. He's interested in the role of art and culture in creating a more resilient and life-affirming world. Many of the themes in this book resonate deeply with him: meditation, homesteading, human-powered travel, being the change, as well as the joys of experimenting together with family and community. These drawings were made while living with his wife, Hiromi, and others at Casa de Paz in East Oakland (Huichin), California.

ABOUT NEW SOCIETY PUBLISHERS

New Society Publishers is an activist, solutions-oriented publisher focused on publishing books for a world of change. Our books offer tips, tools, and insights from leading experts in sustainable building, homesteading, climate change, environment, conscientious commerce, renewable energy, and more—positive solutions for troubled times.

We're proud to hold to the highest environmental and social standards of any publisher in North America. This is why some of our books might cost a little more. We think it's worth it!

- We print all our books in North America, never overseas

- All our books are printed on **100% post-consumer recycled paper**, processed chlorine-free, with low-VOC vegetable-based inks (since 2002)

- Our corporate structure is an innovative employee shareholder agreement, so we're one-third employee-owned (since 2015)

- We're carbon-neutral (since 2006)

- We're certified as a B Corporation (since 2016)

At New Society Publishers, we care deeply about *what* we publish—but also about *how* we do business.

New Society Publishers

ENVIRONMENTAL BENEFITS STATEMENT

For every 5,000 books printed, New Society saves the following resources:[1]

38	Trees
3436	Pounds of Solid Waste
3781	Gallons of Water
4931	Kilowatt Hours of Electricity
6246	Pounds of Greenhouse Gases
27	Pounds of HAPs, VOCs, and AOX Combined
9	Cubic Yards of Landfill Space

[1]Environmental benefits are calculated based on research done by the Environmental Defense Fund and other members of the Paper Task Force who study the environmental impacts of the paper industry.

Certified (B) **Corporation**

FSC
www.fsc.org
MIX
Paper from responsible sources
FSC® C016245

new society
PUBLISHERS
www.newsociety.com